T0259963

AGRICULTURAL RESEARCH COUNCIL
REPORT SERIES
NO. 14

INVESTIGATION OF
VIRUS DISEASES OF BRASSICA CROPS

INVESTIGATION OF
VIRUS DISEASES OF
BRASSICA CROPS

BY

L. BROADBENT

Rothamsted Experimental Station
Harpenden, Hertfordshire

CAMBRIDGE

AT THE UNIVERSITY PRESS

1957

CAMBRIDGE
UNIVERSITY PRESS

University Printing House, Cambridge CB2 8BS, United Kingdom

Cambridge University Press is part of the University of Cambridge.

It furthers the University's mission by disseminating knowledge in the pursuit of
education, learning and research at the highest international levels of excellence.

www.cambridge.org
Information on this title: www.cambridge.org/9781107586758

© Cambridge University Press 1957

First published 1957
First paperback edition 2015

A catalogue record for this publication is available from the British Library

ISBN 978-1-107-58675-8 Paperback

CONTENTS

FOREWORD

From 1948–50 virus diseases caused serious losses of cauliflower and broccoli in most parts of England where these crops are grown intensively. Many field crops in Kent, especially in the Isle of Thanet, failed completely in 1950, as also did broccoli in allotments almost everywhere. The growers became anxious about the future of their industry and the Agricultural Improvement Council asked the Agricultural Research Council to arrange for more research on these diseases aimed to provide practical methods whereby they might be controlled.

Some work on the two main viruses concerned was already in progress in the Plant Pathology Department of Rothamsted Experimental Station, and workers there were approached to see whether this could be extended to a comprehensive study of the conditions leading to serious outbreaks in field crops. This task was undertaken by Dr L. Broadbent, who was particularly well suited for it because of his previous work on the spread and control of virus diseases in potato and lettuce crops.

Of necessity the bulk of the work was done at Rothamsted, but Dr Broadbent found many willing collaborators in the National Agricultural Advisory Service, at Agricultural Colleges and, perhaps most important, among the growers themselves. Consequently experiments and observations were possible in many parts of England, covering a wide range of growing conditions. It is largely a result of this fruitful collaboration that it has been possible to present such a complete account of the many factors affecting the incidence of brassica virus diseases in Britain.

Some results of this work have already been published in scientific journals; these have been included in this report, together with the results of many new experiments by Dr Broadbent or his collaborators. It is thought that the information so assembled will prove valuable both to advisory officers and growers of brassica crops.

1. INTRODUCTION

Virus diseases of cruciferous crops were first recognized in 1921 when Schultz described one in Chinese cabbage, mustard and turnip in the U.S.A. Since then they have been reported from many parts of the world, but they have been most studied in the U.S.A.

Walker, LeBeau & Pound (1945) concluded that the aphid-transmitted viruses of crucifers fell into two groups, which they called cauliflower virus I and turnip virus I groups. Sylvester (1953) agreed with this except that, because of differences in resistance to heat, he thought there might be an intermediate group, which he provisionally called the radish virus I group. The main differences between the two groups are: members of the cauliflower virus I group cause more severe symptoms at temperatures below 24° C. than above; they infect only members of the Cruciferae; *in vitro* they remain infective until heated for 10 min. at temperatures above 70° C.; attempts to make specific antisera against them have failed. Members of the turnip virus I group cause more severe symptoms above 24° C.; they infect plants outside the Cruciferae; their thermal inactivation point is below 70° C., and specific antisera have been prepared against them.

Members of these two groups are common in Britain under the names cauliflower mosaic virus (ClMV) and cabbage black ring spot virus (CBRSV). Smith (1945) recorded that turnips and swedes are sometimes infected with strains of turnip virus I that do not infect cabbage, brussels sprouts and cauliflower; he also found that swedes are occasionally infected with cucumber mosaic virus, which, like ClMV and CBRSV, is transmitted by aphids.

Markham & Smith (1949) described turnip yellow mosaic virus (TYMV), the first virus found affecting crucifers that is not transmitted by aphids, but by biting insects, such as flea-beetles.

Although the properties of the viruses have been studied in considerable detail, few papers have been published on their spread and control in field crops, the most important being those by Caldwell & Prentice (1942, b) and Pound (1946).

This report describes the results of work done, mostly in England, during 1950–5; many aspects of the brassica virus diseases have been investigated, but special attention has been paid to virus spread in the field and possible control measures.

The viruses of watercress (*Nasturtium officinale* R.Br.), an important cruciferous crop in Britain, have not yet been adequately described and are not discussed in this report.

2. BRASSICA CULTIVATION IN ENGLAND AND WALES

Brassica crops are grown both on farms and on horticultural holdings, and are divided for statistical purposes into the agricultural crops: turnip, swede, cabbage, savoy, kohlrabi, kale, rape and mustard, and the horticultural crops: turnip, swede, cabbage, savoy and kale for human consumption, brussels sprouts, summer and winter cauliflower, and sprouting broccoli. The approximate mean acreage under all brassica crops in each county in England and Wales for the years 1947–9, was calculated as a percentage of the total agricultural acreage, and the results plotted in Fig. 1, which thus shows the relative importance of brassica crops in the various counties. The more important brassica-growing areas are in the eastern half of England, in Worcestershire, and in the south-west of England and Wales.

Losses from virus diseases have been greatest in cauliflower, and Fig. 2 shows the counties in which this crop is important. Cauliflower cultivation is usually confined to small areas within counties, so no attempt has been made to place the acreages on a basis proportional to other crops. The biggest acreages of cauliflower are in counties with high proportions of other brassica crops, except in Lancashire and the West Riding of Yorkshire.

Information about the distribution of brassica crops is useful when considering sources of virus and insect vectors, but too much stress must not be laid on statistics, because the incidence of virus diseases depends very much on the proximity of one susceptible crop to another at particular times of the year. Also allotments and gardens provide sources of infection which are large in total area and are widely scattered, although most important in urban areas.

The annual fluctuations in acreages under brassica crops from 1931–54 are shown in Fig. 3. The acreage increased strikingly during the war; the total area under brassicas increased by 44 per cent. from 1939 to 1945, and the total agricultural brassica acreage by 47 per cent.

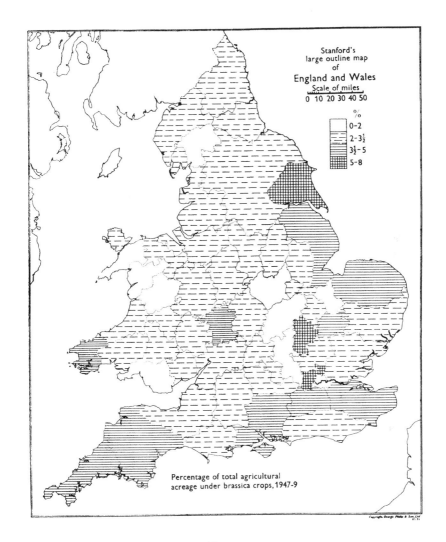

FIG. 1.

3

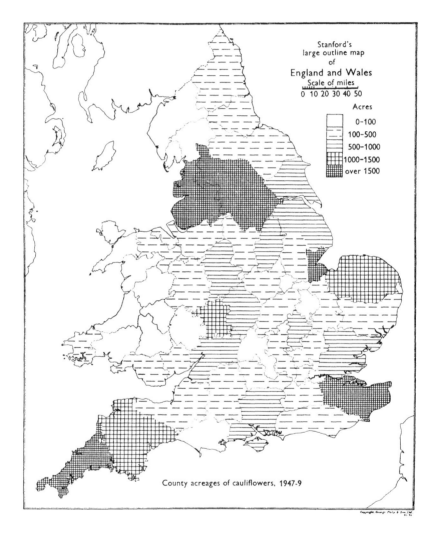

County acreages of cauliflowers, 1947-9

Fig. 2.

4

Horticultural brassica acreages were smallest in 1941 and largest, almost double, in 1948; the cauliflower acreage increased by 106 per cent. between 1940 and 1948. A similar or greater increase in the acreage of brassicas in allotments and gardens occurred over the same period. The

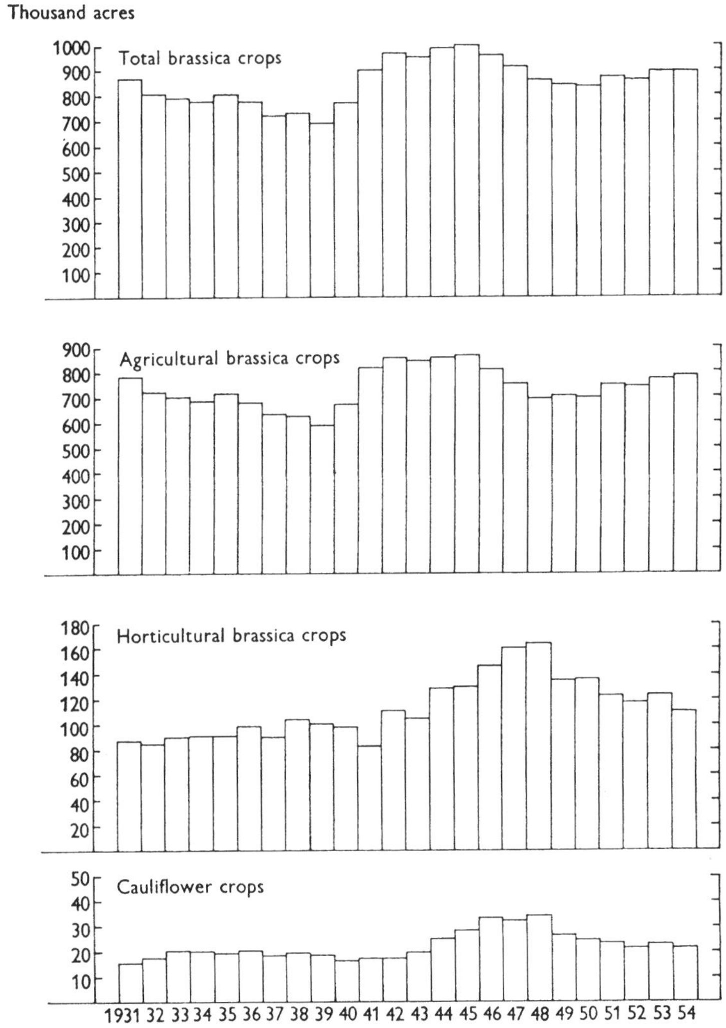

FIG. 3. Annual fluctuations in acreages under brassica crops in England and Wales, 1931 to 1954.

influence of this great increase in brassica growing on virus diseases will be discussed later. There seems little doubt that the 38 per cent. decrease in cauliflower acreage that took place between 1948 and 1952 may be partly ascribed to the severe outbreaks of cauliflower mosaic that occurred during those years.

3. HISTORY OF THE DISEASES IN BRITAIN

Ogilvie (1934) first reported virus-infected plants in brassica crops. Smith (1935), describing CBRS, which he found near Cambridge in the autumn of 1934, said that it probably differed from the mosaic disease (ClM) which was also prevalent in that area. Moore (1943), summarizing the records of advisory plant pathologists, stated that: 'Mosaic has been reported most years since 1934, and from widely separated counties, mainly on cauliflower and broccoli, but also on cabbage, brussels sprouts and kale.' It thus seems likely that both ClM and CBRS had been present in England for some, if not many, years before they were recognized and recorded. Moore further stated that: 'Late in 1942 winter brassicas were severely affected in most allotments and gardens around Harpenden, and Hungry Gap kale was practically killed in several gardens. Symptoms of Black Ringspot were also reported from Cambs. Glam. Worcs. Surrey and Yorks.', and again in 1948 that: 'Both Cauliflower Mosaic and Cabbage Black Ring Spot were exceptionally abundant, and caused much damage in almost all districts in 1944. They were again common, though less destructive, in 1945 and 1946.' Caldwell & Prentice (1942 a) stated that ClM had been observed yearly since 1936 in Devon and Cornwall, where it was widespread, and in some seasons, very serious. Millard (1945) wrote: 'Only during the last few years has it [ClM] become noticeable on broccoli and other brassicas in Yorkshire. The disease is often widespread in the crop, and it is not uncommon to see up to 75 per cent. or more of the plants attacked and ruined.' Kvicala (1948 a) studied the transmission of the two viruses, which were often found together in the same plant near Cambridge.

During 1948–50 ClMV caused severe losses in almost all the cauliflower-growing areas, except Cornwall. During 1950–1 the disease was still prevalent, but, although some crops suffered badly, it was less prevalent than during the previous few years. This improvement continued and during 1951–2, although there was still considerable loss of winter cauliflowers in the midlands and north, the south was almost free. During 1952–3 the incidence of ClM was again moderate in the north and low in the south, but the virus spread rapidly in the midlands and caused much loss. CBRS was also prevalent this year, after being rare for a few years. Both diseases were generally infrequent during the

6

years 1953–5, but caused damage in some places where seedbeds or crops were near sources of virus and aphids.

In Scotland ClM was first recorded in 1940, and CBRS in 1939, in the Forth region (Dennis & Foister, 1941). Both diseases were noted in the Lothians in 1943 and 1944, and caused much loss in 1945. They have also been recorded in Aberdeen, Lanark and Ayrshire. Since 1945 they have become less prevalent, and even very rare since 1952 (Department of Agriculture for Scotland, *in litt.*).

Markham & Smith (1946) first obtained TYMV from turnips growing near Edinburgh. They stated (1949) that it was common in this area, and had also been found near Bristol (in turnips at Bromham, Wilts.— Mr L. Ogilvie, *in litt.*). During 1954 the disease was found in turnips near Dundee and Udny, Aberdeenshire.

Croxall, Gwynne & Broadbent (1953) identified a strain of TYMV as the cause of serious losses in winter cauliflower and other brassica crops in north-east England.

The discovery by the writer in 1953 of another virus disease, turnip crinkle, in Scotland suggests that other undescribed viruses may exist locally in cruciferous weeds and crops.

4. CAULIFLOWER MOSAIC AND CABBAGE BLACK RING SPOT

PROPERTIES OF THE VIRUSES

The physical properties of the two viruses have been determined in Britain by Caldwell & Prentice (1942*a*), Kvicala (1948*a*) and Hamlyn (1952). These properties vary with the host from which the sap is extracted and the concentration of virus in the sap, so the results of tests are variable.

The workers quoted found that ClMV was inactivated by heating for 10 min. at temperatures between 70 and 80° C.; the virus remained infective after dilution to 1 : 2000, but not 1 : 3000; it withstood storage *in vitro* at 22° C. for 7 but not for 8 days.

CBRSV was inactivated by heating for 10 min. at temperatures between 56 and 65° C.; it remained infective at high dilutions, in one instance at 1 : 10,000; it withstood storage *in vitro* at 22° C. for 48 but not for 54 hr.

REACTION OF PLANTS TO INFECTION

Cauliflower mosaic symptoms in cauliflower

The first symptom on a newly infected cauliflower plant is a 'clearing' of the veins, beginning at or near the base of one or more of the youngest leaves, and gradually extending over them. Even the smallest veins appear as fine yellow lines against the green of the lamina (Pl. 1a). These early symptoms usually appear only on leaves not fully expanded; on the largest of such leaves the area of vein-clearing may be limited to a section of the lamina only. Leaves that are fully expanded when the plant becomes infected remain symptomless. The inoculated leaves usually show no symptoms, though under glass discrete chlorotic local lesions may appear at the infection sites, particularly when light intensity is low (Broadbent & Tinsley, 1953).

The rate at which symptoms develop depends on how rapidly plants are growing. Leaves produced after a plant has been infected for some weeks usually show only a transient or no vein-clearing. Instead the main veins become banded with narrow, continuous, dark green areas, the interveinal areas being light green or mottled (Pl. 1b). Later-developed leaves may have light green bands along the veins, with dark green interveinal areas.

Sometimes the mid-rib is curved, or the interveinal areas puckered or severely distorted (Pl. 2a). Leaves with vein-clearing or vein-banding may also develop small, irregular, light brown spots with cell-proliferation, sometimes referred to in American literature as 'enations', which become dry and brittle, and may lead to holes in the lamina (Pl. 1c). Some plants also develop dark necrotic areas along the veins, or spots ('stipple') on the lamina, varying from minute specks to areas 3 mm. in diameter (Pl. 1d). This symptom was first ascribed to ClMV by van Hoof (1952).

Infection stunts growth, especially of young plants, and older leaves become yellow and drop off sooner than those of healthy plants (Pl. 2b). Curds are often small, and infection encourages the development of bracts in the curd. Because the virus checks the growth of the heart leaves, and often makes them bend outwards, the developing curd may be unprotected and become yellow in summer, or frosted in winter. Many plants infected when young die during the winter. Although plants infected when older survive, their mottled leaves surrounding the curd look sick, and soon yellow when cut, so that the curds seem stale and do not sell so readily as healthy ones.

Cabbage black ring spot symptoms in cauliflower

Symptoms in cauliflower consist of chlorotic spots or rings, varying from 2 to 5 mm. in diameter, usually slightly raised and looking like blisters. These spots lack the waxy 'bloom' of the rest of the leaf, and are often more easily seen by transmitted than by reflected light (Pl. 3 *a*) and on the lower than on the upper surfaces of leaves (Pl. 3 *a*, *b*). They are usually few and discrete, but sometimes are so numerous that they merge to form an indistinct mottle. On older leaves they are occasionally necrotic (Pl. 4 *b*, *c*). Infected plants usually first show symptoms on the youngest leaves, but as the plants grow, symptoms become confined to the lower leaves (Pl. 4 *a*).

Factors affecting symptoms

(i) *Host plant.* ClMV produces symptoms in cabbage, brussels sprouts, kale and kohlrabi that are similar to those in cauliflower, but these plants are usually less distorted and stunted. Vein-clearing is usually distinct in young plants, but vein-banding is often very indistinct. Infected plants may be paler than healthy ones, and cabbage and brussels sprouts may show the necrotic stipple (Pl. 2 *d*). Hungry Gap kale, turnip, swede and mustard suffer more than the other brassica species; their leaves are usually much distorted and necrotic, and many infected plants die (Pl. 2 *c*).

CBRSV symptoms are usually more severe in cabbage and brussels sprouts than in cauliflower, and are usually necrotic (Pl. 4 *d*). Symptoms in the kales are similar to those in cauliflower. Turnip, swede and mustard react severely; early symptoms are a coarse vein-clearing and interveinal mottling, with irregular raised patches of dark green tissue among paler areas, giving a blistered appearance. As with ClMV, the younger leaves are often much twisted and stunted, and the plants may die if infected when young. This virus usually kills mustard.

Several brassica species and varieties were grown at Rothamsted in 1951 and 1952 to gain experience of the range of symptoms and to study the spread of the two viruses. In 1951 the plots were planted at the end of June, each with thirty healthy seedlings, five infected with ClMV and five with CBRSV. A similar plan was followed in 1952, except that the plots contained only thirteen healthy plants. Records of symptoms are summarized in Table 1.

Commercial brassica crops are often genetically impure, and symptoms produced by one strain of virus in individual plants of a variety were not constant. In 1951 there was some evidence that different varieties,

9

TABLE 1. Symptom distinctness and plant reaction during summer and autumn in some virus-infected brassicas

	Cauliflower mosaic (vein-clearing and banding)				Cabbage black ring spot
	Early symptoms	Later symptoms	Stunting	Leaf distortion	
	(1)	(2)	(3)	(4)	(5)
Cauliflower (*Brassica oleracea botrytis*)					
1951					
Majestic	++	++	+++	+++	++N
All the Year Round	++	++	+	+	++N
Roscoff 1	++	++	+	++	++
Roscoff 5	++	++	+++	++	++N
St George	++	−	+	+	++
Cambridge Hardy Late	++	−	++	++	+
Purple Sprouting	++	++	++	+	++N
1952					
Early London	++	++	++	+++	++N
Cambridge 5	++	++	++	++	++N
Majestic	++	++	+++	+++	+N
Novo	++	++	++	++	++N
Pioneer	++	++	++	+++	+
Remme	++	++	++	++	++N
Veitch's Self Protecting	++	++N	+	+	++N
Extra Early Roscoff	++	++N	++	+++	++
Morse's November	++	+N	++	++	+
Yuletide	++	−N	++	++	++
Roscoff 1	++	++N	+++	++	++
White Sprouting	++	−−	o	o	+
Roscoff 2	++	++N	+++	++	++
Snow White	++	−N	++	++	++N
Cambridge Dwarf April	++	−−	+	+	+N
Penzance	++	++	+	+	++N
Satisfaction	++	++	+	+	++N
Continuity	++	++	+	+	++N
Late Queen	++	+	+	+	++N
Late Feltham	++	++	++	++	++N
Midsummer	++	++	++	++	+
Cabbage (*Brassica oleracea capitata*)					
1951					
Early Durham	+	−	o	o	++N
Primo	++	−−	o	o	++N
Red Cambridge	+	−−	o	o	++N
Xmas Drumhead	+	−	+	o	++N
January King	++	−	o	o	+
Savoy Drumhead	+	−	o	o	+
1952					
Utility	++	−	o	o	++N
Winnigstadt	++	−N	o	o	++N
Copenhagen Market	+	−−N	o	o	++N
Dutch Savoy	+	−−N	o	o	++N
Early Jersey Wakefield	++	−	o	o	++N
Early Offenham	++	−−	o	o	++N

TABLE I. *Continued.*

	Cauliflower mosaic (vein-clearing and banding)				Cabbage black ring spot
	Early symptoms	Later symptoms	Stunting	Leaf distortion	
	(1)	(2)	(3)	(4)	(5)
Kale (*Brassica oleracea acephala*)					
1951					
Marrowstem	++	+	+	++	++N
1952					
Thousand Headed	++	+	o	o	++N
Variegated	++	--	o	o	+
Extra Curled Scotch	++	--	o	o	++N
Jersey	++	--	o	o	++N
Brussels Sprouts (*Brassica oleracea gemmifera*)					
1951					
Ashwell's strain	+	+	o	o	+
Rous Lench	++	+	o	o	+
Cambridge 1	++	++	+	o	++
1952					
Ashwell's strain	++	--	o	o	++N
Rous Lench	++	+	o	+	++N
Cambridge Special	++	-	o	o	++N
Kohl Rabi (*Brassica oleracea caulorapa*)					
1952	++	-	o	o	++N
Rape (*Brassica napus*)					
1952					
Lembke's Winter	+	+N	+++	+++	++N
Hungry Gap Kale	++	-	+++	+	+
Turnip (*Brassica rapa*)					
1951-2	++	D	+++	+++	++D
Swede (*Brassica napo-brassica*)					
1952	++	++	+++	+++	++N
White Mustard (*Sinapis alba*)					
1951	++	D	+++	+++	++D

Column (1) ++ distinct, + indistinct; (2) ++ distinct, + indistinct, − often masked, − − masked, N necrotic stipple, D plants died; (3) and (4) o none, + little, ++ moderate, +++ much; (5) ++ distinct, + indistinct, N older leaves necrotic, D plants died.

Records in winter cauliflower, 1952–3

	Symptoms ++ distinct + indistinct	Defoliation − little + moderate ++ severe +++ very severe			Plants dead/23
		20 Dec.	21 Jan.	6 March	6 March
Roscoff 2	++	+	+++	+++	10
White Sprouting	+	−	−	+	0
Snow White	++	+	++	+++	4
Cambridge Dwarf April	+	−	−	+	6
Penzance	++	+	++	+++	4
Satisfaction	++	+	+++	+++	8
Continuity	++	−	−	+	2
Late Queen	++	−	+	+	0
Late Feltham	++	++	+++	+++	9
Midsummer	++	−	+	++	2

especially of cauliflower, differed in susceptibility to infection with ClMV. The percentage infections at the end of November were:

Roscoff No. 5	80
Purple sprouting broccoli	54
Roscoff No. 1	48
St George	44
Cambridge Hardy Late	37

Plants not only varied in susceptibility, that is, the ease with which they became infected, but infected plants differed considerably in their ability to tolerate infection; this is illustrated in the columns headed 'Stunting' and 'Leaf distortion' in Table 1. No information on susceptibility to infection was obtained in 1952, for both viruses spread rapidly and most plants of all varieties became infected.

Mr E. Lester also reported that in the Melbourne area of Derbyshire ClMV was very prevalent, and though the varieties Satisfaction and Late Feltham suffered severely, St George was tolerant and Continuity even more so, though all varieties showed definite symptoms.

The National Institute of Agricultural Botany and Rothamsted are now co-operating in the work of testing cauliflower varieties for their susceptibility to infection and their ability to tolerate infection. Differences between existing varieties are large enough to encourage the hope that new ones could be bred which would resist infection or tolerate it much more than those now generally grown.

Differences in susceptibility were found also in a variety trial at Stockbridge House Experimental Horticulture Station, made by Dr I. F. Storey and Mr F. G. Smith in 1951, with six local Yorkshire winter cauliflower strains. The seedlings were mostly raised away from sources of virus and were healthy when transplanted, except Variety 4, which was raised at Wakefield and had some infected seedlings, but these were removed soon after transplanting. Plants were transplanted into four blocks, each containing one plot of 121 plants of each variety; there was an infected plant in the centre of each plot. New infections were recorded weekly.

Fewer plants of Varieties 2 and 6 became infected than of the others (Fig. 4); the virus also spread differently in the plots of these two varieties, and the percentage of infected plants near to the 'infector' was smaller on 18 September.

Many garden flowering plants are susceptible to CBRSV, and some, such as *Cheiranthus cheiri*, *Papaver* spp., *Petunia hybrida*, *Verbena hybrida*, *Zinnia elegans* and *Matthiola incana* are stunted or severely

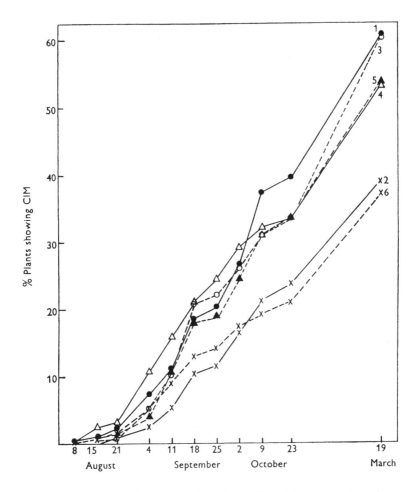

FIG. 4. Percentage increase of ClM in six Yorkshire growers' strains of winter cauliflower, 1951–2.

distorted. The 'broken' flowers of infected wallflowers, stocks and honesty are a common sight; these three species (wallflower rarely) may also be infected with ClMV, which causes severe stunting, but not flower 'break'. Weed hosts, inoculated in the seedling stage, were usually more susceptible and reacted more severely to CBRSV than to ClMV. The weeds *Capsella bursa-pastoris*, *Lepidium campestre*, *Camelina sativa*, *Neslia paniculata*, *Brassica nigra*, and *Sinapis arvensis* were killed by CBRSV.

Susceptibility to infection when inoculated with sap decreases as plants grow older; frequent attempts to infect large cauliflower plants grown outside or under glass, even when 'Celite' was incorporated in the inoculum, all failed. Symptoms of both diseases developed sooner

13

under glass in young than in older seedlings; thus three-leaved cauliflower seedlings inoculated with ClMV all showed symptoms within 23 days, compared with 27 days for six-leaved plants; comparable times for CBRSV were 12 and 16 days. Inoculating young leaves gave more infections and produced symptoms sooner than inoculation of old leaves (Table 2).

TABLE 2. Number of plants infected, and time of appearance of symptoms in batches of eight cauliflower seedlings, when the upper two, or lowest two leaves were inoculated,* 10 December 1951

Days after inoculation	Cauliflower mosaic virus. Leaves inoculated		Cabbage black ring spot virus. Leaves inoculated	
	Upper two	Lowest two	Upper two	Lowest two
12	0	0	1	0
14	1	0	3	0
17	4	1	6	3
21	7	1	6	3

* Inoculation in this report refers to sap-inoculation, not insect, unless specifically stated.

In general, ClMV was much more easily transmitted by inoculation than CBRSV; over the period of a year 93 per cent. of the cauliflower seedlings inoculated with ClMV became infected, compared with 53 per cent. of those inoculated with CBRSV.

Miss Hamlyn, transmitting ClMV with aphids, obtained more infections when using turnip as a source plant than when using cauliflower. Also turnip was more readily infected than cauliflower.

(ii) *Virus strains.* (a) *Cauliflower mosaic virus.* Several isolates of the virus have been obtained from cauliflower crops. They were inoculated to Chinese cabbage and turnip and could be classified into three groups according to the reactions of these plants: (1) severe strains, causing distinct vein-clearing, much leaf distortion, and soon killing; (2) moderate strains, causing distinct vein-clearing, moderate distortion, but not killing; and (3) mild strains, causing less distinct vein-clearing and little or no leaf distortion.

The order of severity towards Chinese cabbage and turnip was the same as towards cauliflower. Symptoms caused by severe strains not only appeared earlier than with mild strains, but more plants became infected (Table 3).

Jenkinson (1955) noted that a severe and a mild strain of ClMV were common in Devon; the severe strain caused vivid symptoms and much reduced plant size; the mild strain, with faint and often evanescent symptoms, had little effect on plant size.

Leaf mottles due to weather conditions or manurial deficiencies can be confused with symptoms of ClMV, especially during the winter.

TABLE 3. Number of plants infected, first appearance and severity of symptoms, in cauliflower seedlings

(a) Of three varieties, inoculated with severe and moderate strains of ClMV, 28 December 1951

Date of appearance	Days after inoculation	Extra Early Roscoff		St George		Majestic		Total	
		Severe	Mod.	Severe	Mod.	Severe	Mod.	Severe	Mod.
				Numbers of plants infected					
16 Jan.	20	2	0	2	0	1	0	5	0
19 Jan.	23	3	0	3	0	2	0	8	0
21 Jan.	25	.	1	.	1	2	0	8	2
24 Jan.	28	.	1	.	3	3	2	9/9	6
28 Jan.	32	.	2	.	.	3	.	.	8/9
				Numbers of plants showing symptoms					
Severe symptoms		3	0	3	1	3	0	9	1
Moderate symptoms		0	1	0	1	0	1	0	3
Mild symptoms		0	1	0	1	0	2	0	4

(b) Of Extra Early Roscoff seedlings inoculated with severe, moderate and mild strains of ClMV, 4 March 1952

Strain	Plants infected	Symptoms on 8 April
Severe	3/3	Distinct vein-clearing and banding.
Moderate	3/3	Moderately distinct vein-clearing; one with distinct, two with indistinct vein-banding.
Mild	1/3	Very indistinct vein-clearing; no vein-banding or other symptoms on young leaves.

Other conditions that may sometimes be confused with virus disease symptoms are produced by the toxic saliva of *Brevicoryne brassicae* (Pl. 5d), and by fungi which kill aphids, causing small dark spots on the leaves at the place where the aphids were feeding (Pl. 5c).

(b) *Cabbage black ring spot virus.* Two very distinctive strains of this virus were tested. The severe strain (from Dr K. M. Smith of Cambridge) invariably produced symptoms more quickly than the mild; for example, two batches of ten Extra Early Roscoff cauliflower seedlings were inoculated, using 'Celite', with one or the other of the two strains. Symptoms appeared as follows:

After 14 days   5 severe   2 mild
After 17 days   7 severe   4 mild
After 21 days   7 severe   6 mild

The mild strain caused a few non-necrotic raised 'blisters', which contrasted with the numerous necrotic 'blisters' and rings caused by the severe strain (Pl. 3a; the mild strain did not show clearly in photographs).

(iii) *Mixed infections.* Plants infected with both viruses were sometimes found in the field. The symptoms shown depended upon the length of time the plants had been infected. When cauliflower plants were infected simultaneously with ClMV and CBRSV, early symptoms of both appeared, but soon were more or less obscured by generalized mottles, in which ClM-type symptoms usually predominated (Pl. 5 a, b). If the plant reacted necrotically to the viruses it was often difficult to distinguish CBRS necrosis from ClM stipple and veinal necrosis. Plants infected with both viruses usually reacted more severely and were more stunted than those infected with either virus alone.

(iv) *Weather*. Temperature not only affects plant growth, but also the development of symptoms in infected plants. Pound & Walker (1945 a, b) found that symptoms of ClM appeared sooner in cabbage plants at 24 or 28° C. than at 16 or 20° C., but that they were less severe than at the lower temperatures. At both 16 and 28° C. symptoms of ClM were more severe, and those of CBRS less severe, at short periods (8 hr.) of daylight than at long ones (15 hr.). CBRS symptoms also developed more rapidly at high temperatures, but they were less severe at low than at high temperatures.

The observations at Rothamsted on infected cauliflower plants grown under glass agree with the conclusions of Pound & Walker; ClM was masked when the plants were grown at temperatures above 24° C. Table 4 shows the average and range of time taken for initial symptoms to appear in Extra Early Roscoff cauliflower seedlings inoculated in the 3–5 leaf stage, at different times of the year during 1951–2. The slow growth during the winter delayed the appearance of ClM symptoms more than those of CBRS.

TABLE 4. Times of appearance of symptoms in glasshouse-grown cauliflower seedlings after inoculation with ClMV and CBRSV

Period of inoculation	Mean temperature ° C.	Cauliflower mosaic virus			Cabbage black ring spot virus		
		No. of plants	No. of days before appearance of symptoms		No. of plants	No. of days before appearance of symptoms	
			Mean	Range		Mean	Range
November–January	17·5	47	24	19–40	31	17	12–35
February–March	20·5	71	23	14–41	98	13	11–28
April–October	24·0	201	17	11–26	69	16	9–25
Mean	21·9	319	19	11–41	198	15	9–35

The sudden development of ClM symptoms in spring, when plants start to grow again, has sometimes been attributed to early spring infection, but aphids are inactive at this time and almost invariably such plants have become infected during the previous autumn. Direct evidence on the time taken for ClM symptoms to develop outside was obtained by infecting plants under nylon cages with ten infective apterous aphids: seedlings were transplanted on 13 June, and on 7 July ten were infested with aphids; only two became infected, and symptoms first showed after 31 and 41 days. Of ten plants infested with aphids on 4 August, three became infected, showing symptoms after 34, 41 and 41 days. St George cauliflower seedlings sap-inoculated on 19 November and transplanted outside failed to show symptoms until 17 April of the following year.

Jenkinson's (1955) data, obtained by fortnightly examinations in Devon, suggest that natural infection in July caused symptoms of ClM

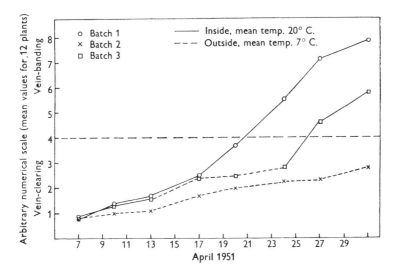

FIG. 5. The effect of temperature on development of ClM symptoms
in cauliflower seedlings.

to show in the transplanted crop in about 8 weeks; infection during August and September showed in about 10 weeks, while late October infections did not show for at least 16 weeks.

Symptoms not only appeared sooner at higher temperatures, but they went through the various phases more quickly when the plants were growing rapidly. Fig. 5 illustrates this effect: forty-eight winter cauliflower seedlings, of a very susceptible Yorkshire variety, were inoculated with ClMV in the 3–5 leaf stage on 13 March 1951, and kept under glass; symptoms had begun to appear on 7 April, when twenty-four were put outside. After one week, twelve of the plants outside were exchanged for twelve of those inside, and after a further 10 days these two batches were again changed. Mean temperatures over the period were 20° C. inside the glasshouse, and 7° C. outside. The succession of symptoms on each leaf of each plant was noted and their type recorded on the following arbitrary numerical scale:

Vein-clearing	Youngest leaf	1
	2nd and 3rd leaves	2
	General	3
Vein-banding	1st leaf	1
	2nd leaf	2
	3rd leaf	3, etc.

The plants which were inside all the time grew more rapidly and vein-banding symptoms appeared within 14 days of their first showing vein-clearing, whereas symptoms on plants outside all the time had developed little in 24 days. Moving the plants inside or outside increased or decreased both the growth of the plants and the extent to which they showed symptoms.

This experiment was followed by one in which ten varieties of cauliflower, infected with ClMV, were grown at two temperatures. Seedlings were inoculated on 14 April, and symptoms began to appear on 28 April. On 3 May each variety was divided into two groups of five plants, with approximately similar symptom ratings; one group was left in the glasshouse, the other was put outside until 22 May. Symptoms were recorded at intervals according to the following arbitrary numerical scale:

1 Vein-clearing on one leaf.
2 Vein-clearing on two leaves.
3 Vein-clearing on three leaves.
4 Vein-clearing on four leaves.
5 Vein-clearing on five to six leaves, or vein-banding on one leaf.
6 Vein-clearing on seven to eight leaves, or vein-banding on two leaves.
7 Vein-clearing on nine to ten leaves, or vein-banding on three leaves.
8 Vein-banding on four leaves.
9 Vein-banding on five leaves.
10 Vein-banding on six leaves.

The mean temperatures over the period 3–22 May were 24° C. inside the glasshouse and 13° C. outside. Again the plants inside grew more rapidly and their symptoms developed more rapidly than those outside. The mean 'scores' for the ten plants of each cauliflower variety on three of the examination dates are given in Table 5. It was as a result of this test that Extra Early Roscoff was chosen for most of the glasshouse work.

TABLE 5. Numerical scores indicating time of appearance and severity of ClM symptoms in ten cauliflower varieties, inoculated 14 April 1951

	30 April	7 May	22 May
Extra Early Roscoff	2·7	5·4	8·1
A Yorkshire variety	1·7	5·6	7·7
Roscoff 5	1·8	4·1	7·6
St George	2·1	4·5	7·4
Purple Sprouting Broccoli	0·9	2·6	7·3
Cambridge 5	0·7	3·3	6·7
Majestic	1·2	2·9	6·0
Elsom's March	1·2	2·6	5·8
Roscoff 1	0·9	2·3	5·3
Cambridge MX	0·9	1·9	5·3

The Extra Early Roscoff and St George varieties were retained after 22 May. Indoors at a mean temperature of 25° C., vein-banding on younger leaves of all the plants became indistinct; the interveinal areas became dark green, without bloom, and the vein-banding became dark metallic blue with bloom. In all plants except one, most new leaves showed transient symptoms only, and symptoms failed to develop on these leaves even when the plants were put outside for a fortnight. All the older leaves retained their symptoms till senescence. Symptoms were not masked on any of the plants outside at the mean temperature of 17° C.

Field temperatures above 24° C. are rare in Britain, except for very short periods, and heat-masking of symptoms has rarely been recorded in growing crops, except with a very mild strain of ClMV in Devon (Jenkinson, 1955). During winter also, the symptoms caused by mild strains of both ClMV and CBRSV in cauliflower often become indistinct, although as with heat-masking, the symptoms are usually retained on the outer leaves, unless completely obscured by frost damage.

The effect of light intensity on symptom development was tested by growing Extra Early Roscoff seedlings under glass with some of the plants under a cage covered with two layers of muslin. The results of two experiments with ClMV were very similar; in one, thirty-two seedlings were divided into four batches after inoculation on 4 October 1951. The number of plants showing symptoms on three subsequent dates was as follows:

	Unshaded	Shaded	Unshaded till 18 October, then shaded	Shaded till 18 October, then unshaded
18 October	5	1	5	0
20 October	7	5	7	8
22 October	7	8	7	8

Vein-clearing was recorded and allotted a numerical score as follows: faint 1, moderate 2, very distinct 3. Mean symptom severity is shown in

Fig. 6. Shading not only delayed the appearance of symptoms, but diminished their severity. Ten of the fifteen plants shaded after inoculation showed chlorotic local lesions on inoculated leaves; none showed on the unshaded plants. Chlorotic local lesions sometimes showed on inoculated leaves in mid-winter when light was poor.

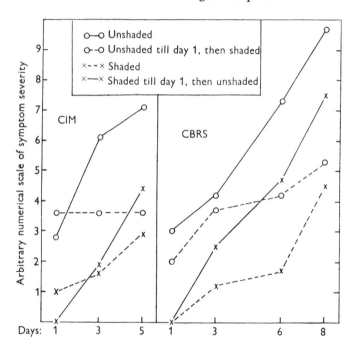

FIG. 6. The effect of light intensity on development of ClM and CBRS symptoms in cauliflower seedlings.

An experiment with four batches of six plants inoculated with CBRSV gave similar results. The plants were inoculated on 23 October 1951, and the numbers of plants showing symptoms on four subsequent dates were as follows:

	Unshaded	Shaded	Unshaded till 3 November, then shaded	Shaded till 3 November, then unshaded
3 November	3	0	3	0
5 November	5	4	3	2
8 November	6	4	6	3
10 November	6	4	6	4

Symptoms were again allotted a severity score of 1, 2 or 3, and the mean severity on the four recording dates is shown in Fig. 6. On 10 November sap from the leaves with symptoms, from each of three plants in each batch, was rubbed on four half-leaves of tobacco per plant. The total numbers of local lesions that developed were: unshaded 122, shaded 391, unshaded, then shaded 353, shaded, then unshaded 114; so the poor symptoms in the shaded plants were not caused by a low concentration of virus.

(v) *Manuring.* Broadbent & Tinsley (1953) found that nitrogen significantly increased the rate at which ClM symptoms developed in Extra Early Roscoff cauliflowers; it also increased their severity, whereas phosphate decreased them and potash had no significant effect. Phosphate decreased symptoms less when potash was given than when it was not.

During 1953–4 and 1954–5 experiments at Rothamsted and Efford Experimental Horticulture Station tested the effects of nitrogenous manuring on the yield and quality of healthy and virus-infected cauliflowers.

ROTHAMSTED, 1953–4

The experiment consisted of five randomized blocks of six plots in an arable field, rich in potash and phosphate and moderately supplied with humus and nitrogen. The previous crop was wheat. The six treatments were:

(1) Nil   No additional nitrogen.
(2) D   Dung at 20 tons per acre.
(3) D 5H   Dung (20 tons)+5 cwt. per acre of hoof and horn meal.
(4) D 10H   Dung (20 tons)+10 cwt. per acre of hoof and horn meal.
(5) 5H   5 cwt. per acre of hoof and horn meal.
(6) 10H   10 cwt. per acre of hoof and horn meal.

The dung was applied and ploughed in on 9 March, the hoof on 12 June. The basal fertilizers, 4 cwt. superphosphate and 2 cwt. muriate of potash per acre, were applied when the cauliflowers, variety St George, were planted on 15 June.

Each plot had twelve rows of twelve plants, 27 in. × 27 in., but the edge rows were rejected and records were taken only on ten rows of ten plants. Eight seedlings infected with ClMV were planted per plot on 6 July. Virus infected plants were recorded at intervals throughout the growing season, and at harvest each curd was weighed, measured and graded for quality.

The plants on the high-nitrogen plots (D, D 5H, D 10H) were considerably larger in the autumn than those on the low-nitrogen plots (Nil, 5H); those on the 10 H plots were intermediate.

This field did not contain an aphid trap, but there was one nearby. *Myzus persicae* were fairly numerous during July (7 trapped), August (15 trapped) and September (5 trapped); *B. brassicae* during August (11 trapped); a few of each were trapped during October and were seen on the plants during October and November.

Disease incidence. The incidence of ClM (Table 6) increased steadily throughout the autumn, and many infected plants showed symptoms for the first time in spring. During late autumn and winter symptoms were masked in many plants.

TABLE 6. Percentage incidence of ClM in the Rothamsted manurial experiment, 1953–4

Treatments

	Nil	D	D5H	D10H	5H	10H	Mean
6 July	8·0	8·1	8·2	8·0	7·9	8·0	8·0
17 August	12·1	14·6	16·2	14·8	12·3	14·7	14·2
7 September	20·7	23·7	26·8	26·0	20·4	26·1	24·0
6 October	33·4	34·7	40·6	39·2	33·8	38·4	36·7
6 November	40·8	44·0	49·2	50·9	43·1	46·4	45·8
29 March *	65·5	68·3	72·6	71·8	67·7	73·6	69·6
Harvest *	77·7	77·8	84·2	80·3	74·2	84·2	79·4
	(345)	(315)	(285)	(238)	(322)	(254)	

* Incidence on 29 March and at harvest (1 April–24 May) is given as the percentage of the plants still living (shown in brackets).

Increasing nitrogen increased the incidence of ClM, and on 6 November 51 per cent. of the plants in the D10H plots showed symptoms, compared with 41 per cent. in the Nil plots.

Death and 'tip-burn'. 'Tip-burn' or 'scorch' is presumed to be a physiological disorder, the cause of which is unknown; the inner leaves developed necrotic edges, and the stem and heart often rotted later. A record of affected plants on 6 November showed a positive correlation with increasing nitrogen (Table 7). Approximately equal numbers of plants infected with ClMV and uninfected showed this condition.

TABLE 7. Numbers of plants showing 'tip-burn' on 6 November; numbers of plants that died; percentages of plants with and without 'tip-burn' that died. Rothamsted manurial experiment, 1953–4

Treatment	Numbers of plants with 'tip-burn'	Numbers of plants dead	Percentage of plants dead	
			With 'tip-burn'	Without 'tip-burn'
Nil	18	152	61·1	29·4
D	50	178	70·0	32·3
D5H	105	215	66·7	36·7
D10H	140	259	75·0	43·1
5H	60	174	45·0	33·7
10H	94	244	75·5	42·8

Deaths (complete death of the plant during the winter, or rotting of the centre during the spring) were also increased by nitrogen (Table 7); an average of 68 per cent. of the 'tip-burn' plants died, compared with 36 per cent. of those which did not show the disorder on 6 November. More plants died on plots with a high level of hoof than on those with dung (Fig. 7).

Of the plants showing ClM symptoms in the autumn, 46 per cent. died, compared with 41 per cent. of those without. It seems that in the variety St George, a high level of nitrogen and 'tip-burn' led to more deaths than did virus infection.

Curd quality. The quality of the curds was recorded at harvest as (a) perfect, (b) bracty, (c) loose, (d) 'ricey', (e) discoloured and (f) rotting, and according to the severity of (b) to (f), as marketable or not, if larger than 'buttons'. Discoloured and rotting curds were not related to treatment, as they were mostly those that happened to be exposed to frost during harvest. The infected plants deliberately placed in the plots on 6 July were omitted from the records, as they were somewhat root-bound when transplanted, and never made much growth.

The earlier plants became infected with ClMV, the fewer perfect curds they produced, and high levels of nitrogen (D5H, D10H, 10H) were also detrimental (Fig. 8). Infection increased bractiness and looseness, but did not affect riceyness. Nitrogen had no obvious effect on looseness or riceyness, but increased bractiness.

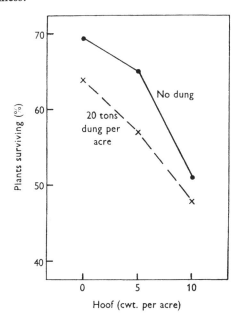

FIG. 7. The effect of nitrogen on plants surviving at harvest, Rothamsted manurial experiment, 1953–4.

Curd size. Curds were put into one of six grades according to diameter, (1) buttons, under 3 in., (2) 3–3¾ in., (3) 3¾–4½ in., (4) 4½–5 in., (5) 5–6 in., (6) over 6 in.

Virus infection significantly decreased curd size, especially of plants that showed symptoms in the autumn. The proportions of 'buttons' in uninfected, spring-symptom and autumn-symptom plants was 2 : 3 : 4; curds over 4½ in. in diameter occurred in the proportions 5 : 4 : 3. The effect of virus infection on mean size of all curds was greater than that of nitrogen, although high nitrogen, especially hoof, resulted in more 'buttons' and a smaller mean size. The low-nitrogen plots produced twice as many large curds (over 4½ in. diameter) as the D10H plots.

Curd weight. Curds were weighed in pounds after being trimmed for market. Virus infection significantly diminished weight, and nitrogen accentuated the effect; with uninfected plants a moderate amount of nitrogen (D, D5H, 5H) increased weight, but larger amounts (D10H, 10H) decreased mean weight to, or below, that of plants in the Nil plots (Fig. 9). The greatest total weight was cut from the Nil and 5H plots; the total weight of the curds in the D10H plots was only 61 per cent. of that in the Nil plots (Fig. 10).

These figures show the effects of treatments on total yields, but as a large proportion of the curds was unmarketable because of failure to meet size or

23

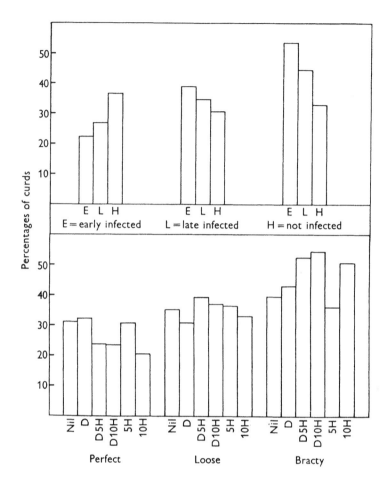

Fig. 8. The effect of ClMV and nitrogen on cauliflower curd quality,
Rothamsted manurial experiment, 1953–4.

quality standards, the figures of commercial interest are those of marketable curds.

Marketable curds. 'Buttons', discoloured, rotting, very bracty, ricey or loose curds were rejected at harvest after they were weighed and their size and quality assessed.

High levels of nitrogen (D 5H, D 10H, 10H) reduced the number of marketable curds; the 10H treatment did so most, and did so even more when alone than when given with dung. This was not only because fewer plants survived in the 10H treatment (Fig. 7), but also because fewer of the curds produced were marketable (Table 8). The same effect of hoof was shown in the total weights of marketable curds, and in the mean weight per curd, which the high level of hoof reduced. These figures probably reflect the diminution in curd size caused by large amounts of nitrogen, especially of hoof (Table 8, percentages of small marketable curds).

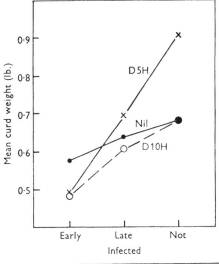

FIG. 9. The effect of ClMV and nitrogen on cauliflower curd weight, Rothamsted manurial experiment, 1953–4.

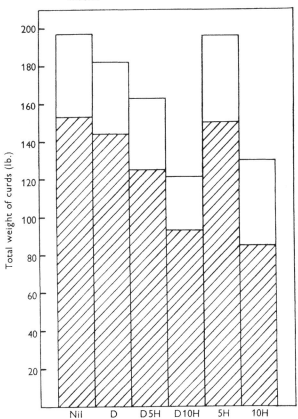

FIG. 10. The effect of nitrogen on total yield and on yield of marketable (shaded) cauliflower curds, Rothamsted manurial experiment, 1953–4.

TABLE 8. Effects of nitrogen on marketable curds, Rothamsted manurial experiment, 1953–4

Dung: tons per acre	Hoof: cwt. per acre			
	None	5	10	Mean
	Numbers of marketable curds: thousands per acre (±0·305)			
None	3·17	2·93	1·93	2·67
20	2·93	2·36	1·98	2·42
Mean (±0·215)	3·05	2·64	1·95	2·55
Difference (±0·431)	−0·24	−0·57	+0·05	−0·25 (±0·249)
	Percentages marketable of total numbers of curds			
None	52·2	52·7	44·3	49·7
20	54·3	48·8	48·5	50·6
Mean	53·3	50·8	46·4	50·1
Difference	+2·1	−3·9	+4·2	+0·9
	Percentages of small marketable curds (of total curds) —diameter 3–4½ in. (± 3·30)			
None	72·5	71·8	84·2	76·2
20	77·2	76·2	81·8	78·4
Mean (±2·34)	74·8	74·0	83·0	77·3
Difference (±4·67)	+4·7	+4·4	−2·4	+2·2 (±2·70)
	Weights of marketable curds: tons per acre (±0·015)			
None	1·17	1·16	0·65	1·00
20	1·11	0·96	0·71	0·93
Mean (±0·075)	1·14	1·06	0·68	0·96
Difference (±0·149)	−0·06	−0·20	+0·06	−0·07 (±0·086)

TABLE 9. Effects of infection with ClMV on marketable curds, Rothamsted manurial experiment, 1953–4

	Infection showing in		Not infected
	Autumn	Spring	
Percentage marketable of total plants	21·7		36·2
Percentage marketable of total curds	42·1	52·4	66·3
Mean weight of marketable curds (lb.)	0·78	0·88	0·91
Mean size of marketable curds (in.)	4·1	4·2	4·3
Percentage of marketable curds in different size grades			
3 –3¾ in.	41·8	35·7	32·9
3¾–4½ in.	37·7	38·8	39·6
4½–5 in.	13·9	16·2	15·4
5 –6 in.	6·6	8·1	10·8
Over 6 in.	0	1·2	1·3

The effects of ClMV on the number, weight and size of marketable curds are given in Table 9. As there was no means of telling what proportion of the plants that died would have shown infection in spring, the percentage of marketable curds from plants not showing symptoms in autumn cannot be subdivided. Autumn infection reduced the proportion of marketable curds obtained from the total number of plants by two-fifths. The proportion of marketable curds which developed was also affected greatly by infection. Not only were fewer marketable curds produced from infected plants, but they were smaller. Plants that showed no symptoms until spring produced curds intermediate in number and size between uninfected plants and those which showed symptoms in autumn.

Time of maturity. Nitrogen and virus infection both slightly shortened the period to maturity. In the treatment with the greatest amount of nitrogen (D 10H) plants matured on average 6 days earlier than in the Nil plots.

The design was the same as at Rothamsted, but the variety Extra Early Roscoff was used; also 24 tons per acre of dung were used instead of 20, and all plots received lime.

The field was old turf, winter-ploughed and spring-limed. The plants grew very vigorously, and it was concluded later that the field was unsuitable for comparing different nitrogen treatments. All the manures were applied on 15 April. Healthy seedlings were planted 19–23 June, and gapped-up on 15 July. The diseased seedlings were transplanted on 6 July.

A flat sticky aphid-trap was operated in the experiment, but aphids were few (5 *M. persicae* and 8 *B. brassicae* caught between 6 July and 28 October).

Disease incidence. Eight per cent. of the plants in each plot were infected when planted; the percentages of plants showing ClM in late October are given in Table 10. ClM was more prevalent on the D 10H plots than on the others.

TABLE 10. Percentage incidence of ClM on 28 October, Efford manurial experiment, 1953

Dung: tons per acre	Hoof: cwt. per acre			Mean
	0	5	10	
0	41·0	42·8	43·6	42·5
24	39·1	42·2	52·8	44·7
Mean	40·1	42·5	48·2	43·6
Difference	−1·9	−0·6	+9·2	+2·2

Death and 'tip-burn'. Many plants died during the autumn and early winter. Few of these had shown ClM, but many showed 'tip-burn'. As at Rothamsted, the incidence of 'tip-burn' and dead plants was increased by nitrogen, although at Efford the differences were not statistically significant. Ninety-three 'blind' plants were recorded, but there was no significant effect of treatment.

Curd quality. Significantly fewer perfect curds were obtained from infected than from uninfected plants. No significant differences were discernible between figures for loose curds (10 per cent.), flat or 'mushroom' curds (22 per cent.), ricey curds (24 per cent.) or bracty curds (21 per cent.), although fewer ricey and bracty curds were obtained from uninfected than from infected plants.

Curd size and weight (excluding late-transplanted plants). Virus infection decreased the size of early- more than of late-infected plants. Mean curd diameters were: 'infectors' 2·4 in., plants showing symptoms in early September 3·9 in., in late October 4·3 in., not infected 4·4 in. Mean curd weights were 2·8 lb. per uninfected curd, 2·6 for late infections, 2·2 for early infections and 0·7 for the 'infectors'.

27

Marketable curds. Manurial treatment had no significant effect on the percentages of curds marketable, although fewer were obtained from plots with hoof. Virus infection significantly decreased marketability, from 91 per cent. of uninfected plants to 85 per cent. of those showing symptoms in late October, to 70 per cent. of those showing in early September and to 20 per cent. of the 'infectors'.

'Gapping-up'. The plants transferred from the seedbed on 15 July to fill the gaps were treated separately because their growth was not comparable with those planted a month earlier. As a means of comparison, the weights and sizes of the plants adjacent or very near to the late-transplants, with the same status of health or disease, have been added. Numbers were too small to compare treatments: eighty late-transplants and eighty adjacent early-transplants yielded:

	Mean weight (lb.)	Mean size (in.)	Percentage marketable
Early	3·0	4·3	99·9
Late	1·0	2·7	42·5

These figures suggest that 'gapping-up' in this variety is of doubtful value.

ROTHAMSTED, 1954–5

The experiment consisted of a 6×6 Latin Square on a moderately fertile field that had carried wheat for the previous three years. The field was limed in spring 1954. The treatments were:

(1) Nil   No additional nitrogen.
(2) D   Dung at 20 tons per acre.
(3) DN_1 Dung (20 tons)+4 cwt. nitrochalk per acre.
(4) DN_2 Dung (20 tons)+8 cwt. nitrochalk per acre.
(5) N_1 4 cwt. nitrochalk per acre.
(6) N_2 8 cwt. nitrochalk per acre.

The dung was ploughed in on 4 June; basal potash and phosphate were applied as in 1953; 2 or 4 cwt./acre of nitrochalk were applied on 12 July, and a further 2 or 4 cwt. as a top-dressing on 17 March 1955. Seedlings were transplanted on 12 July, 'infectors' on 23 July. The variety Continuity was used because it tolerates infection with ClMV and was used as the standard 'control' variety in the N.I.A.B. varietal susceptibility trials.

Plants in the high-nitrogen plots (DN_1, DN_2) were on average slightly larger than those in the N_1 and N_2 plots in the autumn; plants in the D plots were intermediate, and the smallest were in the Nil plots.

Aphids were scarce throughout the summer and autumn; a flat sticky trap in the seedbed, and then in the experiment, caught six *M. persicae*, but no *B. brassicae*. The incidence of ClM increased from 7 per cent. (planted) on 23 July to 9 per cent. on 29 September, 12 per cent. on 22 October, 13 per cent. on 21 December and 19 per cent. on 21 April. Treatment did not affect incidence. 'Tip-burn' was scarce before Christmas, but in April 20 per cent. of the plants showed it, and the DN_2 and N_2 plots had about 4 per cent. more than the Nil, D and N_1 plots. Of ClMV-infected plants 27·3 per cent. showed 'tip-burn', compared with 17·5 per cent. of uninfected.

More plants died on the plots with dung (i.e. 12·4 per cent. on D, DN_1, DN_2) than on those without added nitrogen or with nitrochalk (i.e. 9·8 per cent. on Nil, N_1, N_2), but the differences were not statistically significant. A quarter of the plants died after transplanting and were replaced ('gapped') in late July or during August. Of the plants transplanted on 16 July, 8·3 per cent. died during the winter, compared with 19·5 per cent. of 'gapped' plants. Of plants showing ClM before 21 December, 14·4 per cent. died, compared with 10·6 per cent. of uninfected plants or those first showing ClM in spring.

Curd quality. About 7 per cent. of the curds were recorded as perfect when cut, a proportion constant for all treatments. Early infection significantly increased bractiness (39·5 per cent. of early infected plants, 31·9 per cent. of late-infected, and 27·1 per cent. of uninfected), but the nitrogen treatments had no significant effect. About 65 per cent. of the curds were ricey, but neither treatment nor infection had any significant effect on riceyness or looseness.

Curd size. Fewer buttons (curds under 3 in. diameter) and more large curds (over 4½ in. in diameter) were obtained from originally transplanted plants in the high-nitrogen plots (average percentages 14·1 buttons, 42·2 over 4½ in. in the D, DN_1, DN_2 and N_2 plots, compared with 16·3 buttons and 31·9 over 4½ in. in the Nil and N_1 plots). Mean diameter was greatest in the treatments with dung. Originally transplanted plants infected before 21 December produced 28·2 per cent. buttons and 25·7 per cent. over 4½ in., compared with 12·5 per cent. buttons and 46·6 per cent. over 4½ in. from uninfected and late-infected plants. Originally transplanted plants produced 16·3 per cent. buttons and 39·4 per cent. over 4½ in., compared with 37·8 per cent. buttons and 20·3 per cent. over 4½ in. from gapped plants.

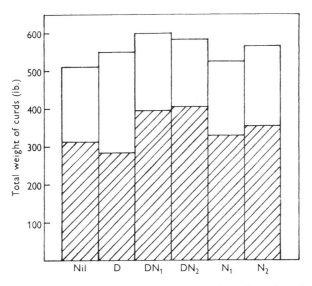

FIG. 11. The effect of nitrogen on total yield and on yield of marketable (shaded) cauliflower curds, Rothamsted manurial experiment, 1954–5.

29

Curd weight. The greatest total weight was cut from the DN$_1$ and DN$_2$ plots (Fig. 11). Nitrochalk significantly increased the total weight of marketable curds, dung alone tended to decrease it (Table 11). Infection with ClMV reduced the mean weight of curds; marketable ones infected early averaged 0·97 lb., late infected 1·26, and uninfected 1·29.

Marketable curds. About 10 per cent. more curds in the D plots ('gapped' plants omitted) were unmarketable than in the other treatments (Table 11). Ten per cent. more curds from uninfected plants or those first showing infection in spring were marketable (51·3 per cent.) than from those showing infection before 21 December (41·3 per cent.). Only 26 per cent. of the 'gapped' plants were marketable, in contrast to 57 per cent. of those originally

TABLE 11. Effect of nitrogen on size, weight and marketability of curds, Rothamsted and Efford manurial experiments, 1954–5

	Dung: tons per acre	Nitrochalk: cwt. per acre			
ROTHAMSTED		None	4	8	Mean
			(±0·222)		
Number of marketable	None	2·66	2·69	2·44	2·60
curds: thousands per acre	20	2·11	3·01	2·69	2·61
Mean	(±0·157)	2·39	2·85	2·57	2·60
Difference	(±0·314)	−0·55	+0·32	+0·25	+0·01(±0·181)
			(±3·26)		
Percentages marketable	None	56·1	56·4	56·2	56·2
of total numbers of curds	20	47·9	61·3	61·6	56·9
Mean	(±2·30)	52·0	58·8	58·9	56·6
Difference	(±4·61)	−8·2	+4·9	+5·4	+0·7 (±2·66)
			(±0·153)		
Weights of marketable	None	1·62	1·70	1·84	1·72
curds: tons per acre	20	1·48	2·05	2·10	1·88
Mean	(±0·108)	1·55	1·88	1·97	1·80
Difference	(±0·217)	−0·14	+0·35	+0·26	+0·16 (±0·125)
EFFORD			(±2·58)		
Percentages marketable	None	62·2	67·8	67·5	65·8
of total numbers of curds	24	69·2	73·0	70·1	70·8
Mean	(±1·82)	65·7	70·4	68·7	68·3
Difference	(±3·64)	+7·0	+5·2	+2·6	+5·0 (±2·10)
Percentages of market-	None	46·2	58·7	53·3	52·7
able curds from plants	24	42·7	55·1	50·7	49·5
showing ClM before early September					
Mean		44·5	56·9	52·0	51·1
Difference		−3·5	−3·6	−2·6	−3·2
			(±0·087)		
Mean curd weight (lb.)	None	1·98	2·24	2·18	2·14
	24	2·34	2·45	2·44	2·41
Mean	(±0·062)	2·16	2·35	2·31	2·27
Difference	(±0·123)	+0·36	+0·21	+0·26	+0·27 (±0·071)

planted. Gapped plants benefited from more nitrogen, 19·1 per cent. being marketable in the Nil, D and N_1 plots, and 32·4 per cent. in the DN_1, DN_2 and N_2 plots.

EFFORD, 1954

The design was the same as at Rothamsted, except that 24 tons of dung per acre were applied instead of 20. Healthy seedlings (variety Extra Early Roscoff) were transplanted on 1 July and the 'infectors' on 7 July.

The crop matured very early and was cut between 6 October and 12 November.

About 5 per cent. of the plants failed and were replaced on 30 July; of these only one out of twenty produced marketable curds, compared with thirteen out of twenty of those transplanted on 1 July.

Aphids were scarce; two *B. brassicae*, but no *M. persicae*, were caught on a flat sticky trap between 1 June and 15 November. The incidence of ClM increased from 8 per cent. on 7 July to 25 per cent. on 5 October. Treatment did not affect incidence.

Treatment had no significant effect on the proportion of perfect (14 per cent.), loose (27 per cent.), ricey (50 per cent.) or bracty (14 per cent.) curds, although there was slightly more bractiness in the high nitrogen plots.

More marketable curds were cut from plots with added nitrogen than from the Nil plots (Table 11), mainly because fewer buttons occurred in these treatments. Dung increased the number of marketable curds and their size and weight, more than did nitrochalk, but among plants infected early with ClMV, nitrochalk increased the percentage of marketable curds more than dung (Table 11). Plots with dung plus 4 cwt. nitrochalk gave the greatest yield. ClMV decreased mean weight (lb.) from 2·38 in uninfected plants to 2·37 in those showing infection on 5 October, 1·84 in those showing infection in early September, and 1·58 in 'infectors'. ClMV infection reduced the percentage of marketable curds from 65 in the uninfected or late-infected plants, to 51 in those showing infection early in September, and to 34 in the 'infectors'.

SUMMARY OF RESULTS FROM THE FOUR MANURIAL EXPERIMENTS

When cauliflower mosaic was prevalent its incidence was increased by increasing nitrogen. Infection with ClMV decreased the size, weight and quality of curds, so that fewer were marketable than from uninfected plants; the earlier that plants were infected, the greater was the loss. Bractiness was the only quality defect to be significantly increased by infection.

High levels of nitrogen, especially of the organic manures dung and hoof, increased the incidence of 'tip-burn'. More plants died during the winter because of high nitrogen than because of ClMV infection. Other varieties might behave differently, for the varieties St George and Continuity tolerate ClMV better than most. Nitrogen had little effect on quality, but moderate amounts increased size and weight. High nitrogen, especially given as hoof or dung, decreased yield, but dung and nitrochalk together increased yield; as a result, fewer curds were marketable from high-nitrogen (especially hoof) plots.

Seedlings transplanted to replace failures a few weeks after the crop was transplanted usually failed to produce marketable curds, and 'gapping-up' seems not to be worth while.

Virus movement and distribution within the plant

The movement of virus into different parts of the plants was studied in the glasshouse. CBRSV causes necrotic local lesions on tobacco leaves and so its concentration in different parts of the plant could be judged, but there is no such quantitative test for ClMV.

(i) *Cauliflower mosaic virus.* Table 12 shows when individual leaves showed symptoms after the two lowest leaves of a three-leaved seedling were inoculated. When symptoms first appeared, all leaves showing symptoms, and those inoculated, contained virus, but not the symptomless ones immediately above the inoculated leaves (Table 13).

TABLE 12. ClM symptom development in a cauliflower seedling inoculated 14 April 1951; leaves numbered from the base of the plant

I, inoculated leaf; O, mottle; C, vein-clearing; B, vein-banding; N, necrotic 'enations' D, leaf dead; — no symptoms.)

Leaf	1	2	3	4	5	6	7	8	9	10	11	12	13	14	15	16	17	18
3 May	I	I	—	—	O¼C	CN	CN	B	.	.	.	.	.	.	.	.	.	.
11 May	D	I	—	—	O	CN	BN	B	B	B	—	.	.	.	.	.	.	.
18 May	.	I	—	—	O	CN	BN	B	B	B	B	B	—	.	.	.	.	.
28 May	.	D	D	—	O	CN	BN	O	B	B	B	B	B	B	C	C	.	.
4 June	.	.	.	D	D	D	D	D	B	B	B	—	—	B	B	B	B	—

TABLE 13. ClM symptoms developed in cauliflower seedlings in which leaves at different levels were inoculated

No. of leaves on plant	Leaves inoculated (numbered from oldest upwards)	1	2	3	4	5	6	7	8	9	10	11	12
5	4 and 5	—	—	—	I	I	—	¼C	C	C	Heat-masked		
5	3 and 4	—	—	I	I	—	—	—	C	C	B	C	Masked
6	3 and 4	—	—	I	I	—	—	—	—	C	C	C	C
6	2 and 3	—	I	I	—	—	—	—	—	—	—	C	C

To find out how soon virus moved out of inoculated leaves, these leaves were removed at various intervals after inoculation. When plants were growing fast, ClMV left inoculated leaves in 5 to 6 days, and symptoms appeared on young leaves in about 16 days; when plants were growing slowly the virus took 7 to 8 days to leave inoculated leaves, and symptoms appeared in about 23 days.

Severin & Tompkins (1948) found that aphids could recover ClMV from plants 6 days after infection when symptoms appeared after 18 days, and 9 to 10 days after when symptoms appeared after 20 to 30 days.

Mature leaves of cauliflower plants in the field often show symptoms on one half only; these are presumably leaves that have not fully developed at the time of infection; virus sometimes occurs in symptomless parts of leaves. Curds and flowers of infected plants contain virus.

(ii) *Cabbage black ring spot virus.* This virus moved out of inoculated leaves in 4 to 5 days when plants were growing fast, and such plants showed systemic symptoms about 9 days after inoculation. When the two youngest or two oldest leaves of five- to six-leaved seedlings were inoculated, and later all leaves above and below these were tested by inoculation to tobacco, all leaves above, but not below, the inoculated youngest leaves contained virus; none of the three leaves immediately above the inoculated oldest leaves contained virus.

Local lesion tests on tobacco showed that CBRSV and ClMV were differently distributed in field plants (Broadbent, 1954). CBRSV first moved from the inoculated leaf to the growing point and later entered some of the subsequently developed leaves, but unlike ClMV, it became concentrated in older leaves as the plant grew, and was often undetectable in young ones (Table 14).

TABLE 14. CBRS symptoms on leaves of a cauliflower plant, and numbers of lesions on tobacco inoculated with sap from them

Leaf	Symptoms	Total lesions
1 (Oldest mature leaf above inoculated leaf)	None	0
2	Very many blisters	955
3	Many blisters	959
4	Many blisters	751
5	Few blisters	419
6	None	29
7	None	40
8	None	8
9 (Youngest)	None	0

CBRSV is not only unequally distributed in leaves of different ages, but also within individual leaves. More virus occurs in the blisters or rings than in symptomless areas of the leaves. No lesions were obtained on tobacco with sap from cauliflower curds or flowers of infected plants. CBRSV concentration increases for some weeks after infection, and then decreases as the plant ages.

The distribution of CBRSV in plants infected also with ClMV was similar to that in plants infected with CBRSV alone. ClM symptoms were usually modified, however, even on the upper leaves, in plants with double infections.

Host range

The following host range list is not intended to be comprehensive, but includes the common crop and weed hosts found in Britain. It is based on the lists given by Caldwell & Prentice (1942a), Walker, LeBeau & Pound (1945), McLean & Cowin (1952–3), and Hamlyn (1952); it also includes species tested by Dr K. M. Smith and Miss

M. E. Short (A.R.C. Virus Research Unit, Cambridge) and the author. Species that have been confirmed as hosts in Britain are marked with an asterisk.

Common British cruciferous crop plants susceptible to both viruses:

*Brassica napus L.	Rape
*B. napo-brassica Mill.	Swede
*B. nigra L.	Black mustard
*B. oleracea L. var. acephala DC.	Kale (hearting, Scotch, ornamental, perennial, tree, sprouting)
*B. oleracea L. var. botrytis L.	Cauliflower (summer & winter)
*B. oleracea L. var. bullata L.	Savoy
*B. oleracea L. var. capitata L.	Cabbage (green & red)
*B. oleracea L. var. caulorapa Pasq.	Kohlrabi
*B. oleracea L. var. gemmifera Zenker	Brussels sprouts
*B. oleracea L. var. italica Plenck.	Sprouting broccoli
*B. rapa L.	Turnip
*Lepidium sativum L.	Garden cress
*Raphanus sativus L.	Radish
*Sinapis alba L.	White mustard

The following cultivated species and weeds are also hosts of both viruses:

*Brassica campestris L.	Wild yellow mustard
*B. chinensis L.	Chinese cabbage
*B. juncea Czern. & Coss	Indian mustard
*Camelina sativa Crantz	Gold of pleasure
*Capsella bursa-pastoris Medic.	Shepherd's purse
*Cheiranthus cheiri L.	Wallflower
*Diplotaxis muralis DC.	Wall rocket
*Iberis amara L.	Candytuft
*Isatis tinctoria L.	Woad
*Lepidium campestre R.Br.	Pepperwort
*Lunaria annua L.	Honesty
*Matthiola incana R. Br. var. annua L.	Annual stock
*Raphanus raphanistrum L.	Wild radish
*Sinapis arvensis L.	Charlock
*Sisymbrium officinale Scop.	Hedge mustard
*Thlaspi arvense L.	Field penny-cress

In addition to the above, the following have been recorded as hosts of cauliflower mosaic virus:

*Armoracia rusticana Gaertn., Mey. & Scherb.	Horse-radish
*Barbarea verna Aschers	Land-cress
*Crambe maritima L.	Seakale
*Diplotaxis tenuifolia DC.	Perennial wall rocket
Hirschfeldia incana Lagreze-Fossat	Hoary mustard

34

In addition to the above, the following have been recorded as hosts of cabbage black ring spot virus:

Abutilon theophrasti Medic.	Velvet leaf
Arabis alpina L.	Alpine rock-cress
Barbarea vulgaris R. Br.	Winter cress
Berteroa incana DC.	Hoary alyssum
Beta vulgaris L.	Sugar beet
B. vulgaris L. var. *cicla* L.	Swiss chard
Calceolaria spp.	
Calendula officinalis L.	Pot marigold
Celosia cristata L.	Cockscomb
Centaurea moschata L.	Sweet sultan
Chenopodium album L.	Fat hen
C. amaranticolor Coste & Reyn.	Red chenopodium
C. glaucum L.	Glaucous goosefoot
Cichorium endivia L.	Endive
Conringia orientalis Dum.	Hare's-ear cabbage
Delphinium ajacis L.	Larkspur
Digitalis purpurea L.	Foxglove
Erysimum cheiranthoides L.	Treacle mustard
Helianthus annuus L.	Sunflower
Hesperis matronalis L.	Dame's violet
Lobularia maritima Desv.	Sweet alison
Lycium halimifolium Mill.	Duke of Argyll's tea-plant
Malcolmia maritima R. Br.	Virginian stock
Matthiola bicornis DC.	Evening scented stock
M. incana R. Br.	Stock
Myosotis palustris L.	Forget-me-not
Nasturtium officinale R. Br.	Watercress
Neslia paniculata Desv.	
Nicotiana glutinosa L.	
N. rustica L.	Aztec tobacco
N. tabacum L.	Tobacco
Papaver nudicaule L.	Iceland poppy
P. rhoeas L.	Shirley poppy
P. somniferum L.	Opium poppy
Petunia hybrida Vilm.	Petunia
Physalis pubescens L.	Barbados gooseberry
Reseda odorata L.	Mignonette
Salpiglossus sinuata Ruiz. & Pav.	Salpiglossus
Scabiosa atropurpurea L.	Sweet scabious
Senecio cruentus DC.	Cineraria
Spinacia oleracea L.	Spinach
Stellaria media Vill.	Chickweed
Tetragonia expansa Murr.	New Zealand spinach
Verbena hybrida Voss.	Verbena
Vinca minor L.	Lesser periwinkle
Zinnia elegans Jacq.	Zinnia

Species

Many species of aphids can transmit one or both viruses (Table 15), but the evidence from field experiments described below suggests that species other than *Myzus persicae* (Sulzer) and *Brevicoryne brassicae* (L.) play little part in the field spread. This was confirmed by exposing batches of potted healthy cauliflower plants. During 1950 and 1951 batches of plants were exposed on bare ground, about 30 yd. from

TABLE 15. Aphid species tested as vectors of ClMV and CBRSV

	Recorded or tested by
Vectors of ClMV and CBRSV	
Macrosiphum euphorbiae (Thos.)	B. & H.
M. rosae (L.)	B. & H.
Acyrthosiphon pisum (Harris)	S. & T.; B. & H.
Megoura viciae (Kltb.)	B. & H.
Myzus persicae (Sulzer)	T.; S.; B. & H.
M. ascalonicus Donc.	B. & H.
Aulacorthum circumflexum (Buckt.)	S. & T.; B. & H.
Brevicoryne brassicae (L.)	T.; K.; B. & H.
Aphis fabae Scop.	B. & H.
Sappaphis radicicola (Mordv.)	B. & H.
Vectors of ClMV only	
Dactynotus sonchi (Wlk.)	B. & H.
Nasonovia ribis-nigri (Mosley)	B. & H.
Hyperomyzus lactucae (L.)	B. & H.
Cryptomyzus ribis (L.)	B. & H.
Myzus ornatus Laing	K.
Sappaphis mali (Ferr.)	B. & H.
Cavariella capreae (Fabr.)	
? =*aegopodii* (Scop.)	S. & T.
Hyadaphis sii (Koch)	
?=*foeniculi* (Pass.)	S. & T.
H. mellifera (Hottes)	
?=*foeniculi* (Pass.)	S.
Lipaphis pseudobrassicae (Davis)	S. & T. ⎫ Not tested
Aphis gossypii Glover	S. & T. ⎬ for CBRSV
A. apii Theo.	S. & T.
A. apigraveolens Essig	S. & T.
A. middletoni (Thomas)	
?=*maidi-radicis* Forbes	S. & T.
Sappaphis inculta (Wlk.)	S. & T. ⎭
Vector of CBRSV	
Aulacorthum solani (Kalt.)	K. Not tested for ClM
Species which failed to transmit virus	
Macrosiphoniella sanborni (Gill.)	B. & H.
Cavariella theobaldi (Gill. & B.)	B. & H.
Hyalopterus pruni (Geoffr.)	B. & H.
Coloradoa rufomaculata (Wils.)	B. & H.
Brachycaudus helichrysi (Kltb.)	B. & H.
Tuberolachnus saligna (Gmel.)	B. & H.
Aphis rumicis L.	S. & T. Not tested for CBRSV
Drepanosiphum platanoides (Schr.)	B. & H. ⎫ Not tested
Periphyllus testudinatus (Thom.)	B. & H. ⎬ for ClMV

B. & H., Broadbent & Heathcote (Rothamsted); K., Kvicala (1945, 1948 *b*); S., Smith (1937); S. & T., Severin & Tompkins (1948); T., Tompkins (1937).

allotments carrying virus-infected plants; during 1952 the exposed plants were surrounded at a distance of 5 yd. by cauliflower plants, alternately infected with ClMV and CBRSV. An aphid trap in the centre of the batch of exposed plants at plant height was changed weekly. Although other aphids were often numerous, infections occurred only when *M. persicae* or *B. brassicae* were active (Table 16).

TABLE 16. Aphid trap catches and numbers of plants that became infected when batches of cauliflower plants were put outside for short periods

| Date | Aphids trapped | | | No. of plants put out | No. of plants infected with | | |
	M. persicae	*B. brassicae*	Others		ClMV	CBRSV	Both
		No source of virus within 30 yd.					
24 Aug.–14 Nov. 1950 (4 periods)	0*	1	78	157	0	0	0
15 March–27 June 1951 (5 periods)	0*	0	370	200	0	0	0
27 June–19 July 1951	0*	0*	697	40	0	1	0
19 July–8 Aug. 1951	6	0*	432	29	2	0	0
8–30 Aug. 1951	0	2	110	28	0	0	0
30 Aug.–10 Oct. 1951	1	8	94	24	5	0	1
10 Oct.–22 Nov. 1951	1	1	119	30	3	0	0
		Surrounded by infected plants at a distance of 5 yd.					
17 April–12 June 1952 (2 periods)	2	0	487	60	0	0	0
12 June–10 July 1952	9	4	941	27	3	1	3
10–24 July 1952	1	8	221	28	13	4	5
24 July–14 Aug. 1952	4	86	147	25	13	0	11
14 Aug.–11 Sept. 1952	9	156	38	30	4	0	26
11 Sept.–27 Nov. 1952 (2 periods)	0*	2	41	60	5	1	1

* A few aphids found on the plants, though none caught on the trap.

Severin & Tompkins (1948), Kvicala (1948a), Hamlyn (1952, 1955) and van Hoof (1954) attempted to find out if the viruses were transmitted more readily by *M. persicae* than by *B. brassicae*, or if the aphids transmitted CBRSV more readily than ClMV, but no definite conclusions can be drawn from their results. Under similar test conditions,

Hamlyn found that both aphids transmitted CBRSV to more plants than ClMV; Severin & Tompkins and van Hoof found that *B. brassicae* infected more plants with ClMV than *M. persicae*.

Factors affecting transmission

With many 'non-persistent' viruses, such as potato virus Y, a period of pre-infection fasting and a short infection-feeding time by the aphids increases the number of plants infected; the aphids also retain virus longer if kept without food after the infection-feed than if they are allowed to feed on healthy plants (Watson & Roberts, 1939). The authors mentioned in the previous paragraph studied the relationships of the two viruses and their vectors, and their results indicate that whereas CBRSV is a typical non-persistent virus, ClMV is not typical.

From these results it can be postulated that a winged aphid, either *B. brassicae* or *M. persicae*, would transmit CBRSV most readily after a period of flight, or of absence from a food plant, of at least 15 min. Ten seconds on an infected plant would be sufficient to enable the aphid to pick up virus, but prolonged feeding on the infected plant (i.e. over 15 min.) would considerably reduce its infectivity, unless it were constantly moving and probing the leaf. Having acquired virus, the aphid would transmit it best if it moved to a healthy plant immediately and fed on that plant for some minutes, though it might sometimes transmit virus in as little as 5 sec. If it moved from plant to plant it might transmit virus to many of the plants it visited during half an hour. After leaving the source plant it might still be able to transmit virus to the first plant it fed on during the next 9 hr.

With ClMV the story would be somewhat different. Fasting before feeding on the virus-infected plant would not always increase an aphid's capacity to acquire virus, and it would matter little if it fed for a long time on the source. On moving to a healthy plant, it would transmit virus best if it fed for at least 2 min.; the virus could survive for at least 3 hr. in the feeding aphid, so that it might transmit virus to many plants in that time. After leaving the infected plant, *M. persicae* might still be able to transmit virus to the first plant it encountered during the next 6 hr., but fasting *B. brassicae* could retain the virus for 24 hr.

Incidence of aphids and disease

M. persicae can be an important vector in the field, as is shown by the experiments in Kent and Yorkshire in 1951 (page 65), when *B. brassicae* were absent, but the evidence of recent years suggests that serious outbreaks of ClM are more often associated with movements of many *B. brassicae*. This aphid is often very numerous in late summer, when it leaves one brassica crop for another, often taking virus with it.

Its numbers are usually maximal in August and September, and those of *M. persicae* in July.

Heathcote (1955) studied the populations of *M. persicae* and *B. bras-sicae* that developed on various crop host plants, and concluded that *M. persicae* flourished best during the summer on turnip and mustard, with potato, kale, cauliflower, cabbage and brussels sprouts as other favoured hosts. Lettuce, spinach and sugar beet were colonized, but only small populations developed. *B. brassicae* flourished best on brussels sprouts, with cauliflower and cabbage as other favoured hosts; populations were small on mustard and turnip.

Trap catches and field observations have shown that *M. persicae* are usually more numerous during the spring than *B. brassicae*, because they overwinter viviparously on many herbaceous hosts, and large numbers often develop on lettuce and other crops during April and May; thus *M. persicae* is probably the main agent for the transfer of virus from overwintering crops to seedbeds.

Virus spread within the crop depends largely upon the number and activity of winged aphids, especially those colonizing the crop. Apterous aphids move from plant to plant, and may be responsible for some of the infections that occur in plants adjacent to infected plants (page 72), although the frequency with which such plants escape infection suggests that apterae play little part in virus transmission. Aphids leaving a crop tend to fly upwards and away, so they will usually cause little trans-mission in the crop they are leaving; on the other hand, winged aphids bred on the crop may make abortive or short flights from plant to plant in the evening or in bad weather, and transmit virus at these times.

The aphid traps used from 1950 to 1952 were cylindrical sticky traps (Broadbent, Doncaster, Hull & Watson, 1948), standing with the base $3\frac{1}{2}$–4 ft. from the ground. Aphids are caught on these mainly by wind impaction, either as they leave or enter the crop, or as they are blown across it. In windy weather the number of aphids caught may have little relation to those active in the crop, although the spread of viruses in potato crops has been significantly correlated with the numbers of *M. persicae* caught on this type of trap (Broadbent, 1950). For virus disease studies a horizontal trap, level with the top of the crop, should be better, because aphids landing thereon could have landed on a plant. The water trap of Moericke (1951) is very useful for this purpose, but requires changing frequently, so a yellow (attractive to aphids) hori-zontal sticky trap, to be changed weekly, was devised at Rothamsted, and has been used since 1952 in the brassica virus work (Pl. 6*b*). Heathcote (1955) compared the catches on these different types of traps.

Both *M. persicae* and *B. brassicae* were exceptionally abundant in

Kent and Sussex in 1947 and again in 1949, and this abundance was undoubtedly the cause of the 1948–9 outbreak of ClM (Glasscock and Moreton, 1955). The N.A.A.S. survey of *B. brassicae* on brussels sprouts also showed that this aphid was particularly numerous in England during the years 1947–9 (Empson, 1952).

The great increase in acreage under cauliflowers and other horticultural brassica crops during the years 1946–8 not only provided more plants for aphids to breed on, but also more sources of virus in relatively circumscribed areas.

TABLE 17. Numbers of *M. persicae* and *B. brassicae* caught on traps in cauliflower crops during the period July to October during the years 1951–4 in different parts of England

	M. persicae				*B. brassicae*			
	(Cylindrical traps)		(Flat traps)		(Cylindrical traps)		(Flat traps)	
	1951	1952	1953	1954	1951	1952	1953	1954
Northumberland	.	.	30	2	.	.	1	0
Durham	.	.	34	89	.	.	0	1
Yorkshire	23	34	120	21	0	135	70	2
Lincolnshire (south)	.	.	284*	12*/37*	.	.	20*	1*/4*
Derbyshire	.	.	9	8	.	.	84	2
Staffordshire	.	.	.	2	.	.	.	0
Warwickshire	26	24*	25	2†	60	421*	39	0†
Northamptonshire	.	.	44	.	.	.	71	.
Norfolk	.	.	10*	48	.	.	50*	5
Hertfordshire	17	28	28	6	9	426	13	0
Buckinghamshire	21	10	.	.	55	107	.	.
Berkshire	.	.	.	1	.	.	.	0
Kent	58	5*	7†	6/13	0	39*	0†	1/5
Hampshire	.	11	5	0	.	5	8	2
Somerset	.	.	5	6	.	.	11	3
Devon	27	24	2†	7	20	100	14†	5
Cornwall	.	.	.	1	.	.	.	0

* Not including July.
† Not including September and October.
/ Two-trap sites.

The variation in aphid numbers (Table 17) and virus disease incidence has been considerable since 1950. In the autumn of 1950 ClM was still widespread in all areas of cauliflower production, except Cornwall. During 1951 populations of *M. persicae* and *B. brassicae* were moderate. *B. brassicae* was scarce in some areas of the north and south, but was fairly numerous in the midlands. Virus was transmitted from crop to crop and to seedbeds in the spring, and disease incidence in winter cauliflowers was moderate in the midlands and north, but was low in the south.

During 1952 populations of *M. persicae* were moderate, and did not vary much in different parts of the country, but very large populations of *B. brassicae* developed in a central belt across England, and in this area almost every plant in many cauliflower and other brassica crops was infected. Disease incidence was again moderate in the north and low in the south. CBRSV appeared in all parts of the country during 1952, after having been very scarce for the previous few years. Although it was never so prevalent as ClM, it occurred together with ClM in most places, which suggests that many sources of infection might have been carrying both viruses. As most crops the previous year were infected with ClMV only, it is possible that these sources were perennial crucifers in gardens.

Aphid incidence varied yet again in 1953; both *M. persicae* and *B. brassicae* were very scarce in spring in all areas, and despite the many sources of virus in overwintering crops, especially in the midlands, there was little or no transmission of virus to seedbeds and spring crops. *B. brassicae* populations were moderate in the midlands and Yorkshire later in the summer, but were small further north and in parts of the south; but *M. persicae* populations, though small in the south and west, were very large in the eastern midlands and Yorkshire, and in the few places where virus was present in the crops, considerable spread occurred. During 1954 *M. persicae* were again scarce in spring, but large numbers developed later on potatoes in eastern England. Numbers on brassicas were small, however, and *B. brassicae* were very scarce in all parts of the country throughout the year. Incidence of the brassica virus diseases was everywhere low.

The fluctuations in aphid numbers, which have a considerable influence on virus disease incidence, are brought about by a complex of factors. The number of aphids overwintering is very important because they, and their immediate progeny, transmit virus from over-wintering crops to seedbeds and newly planted spring brassicas.

M. persicae overwinters in the egg stage on peach and some other species of *Prunus*, but winged aphids do not migrate from these hosts until May or June. This method of overwintering is of secondary importance in England, partly because *Prunus* hosts are relatively few, but mainly because living *M. persicae* overwinter readily on numerous hosts, such as lettuce, spinach, brassicas and weeds, in all but the most severe winters, and under shelter every winter in glasshouses, frames and mangold clamps. Winged aphids from these sites migrate early in spring, even in March if the weather is mild (Broadbent & Heathcote, 1955).

B. brassicae overwinters mainly as eggs in the north and east, and mainly as living aphids in the south and west (Empson, 1952). Both eggs and aphids winter on brassica plants. Markkula (1953), in a detailed study of *B. brassicae*, reported that the majority occurred on cultivated crucifers. Multiplication was greatest on two- to three-month-old plants,

41

and progressively lessened on younger and older plants. In seeding plants, the aphids multiplied more rapidly on the flower and fruit stems than on the leaves. The production of young was fairly constant when the plant was well-watered, but inadequate watering led to at least a doubling of the number of young produced; also the less water the plant received, the greater was the proportion of winged aphids produced.

Aphids are usually most numerous in warm, dry weather, and the number in autumn that may lay eggs or overwinter often depends on the weather in late summer. Parasites and predators are often responsible for great fluctuations in aphid numbers. They multiply in seasons when aphids are very numerous in the summer, and often destroy the infestation; many overwinter and prevent infestations of aphids from developing next spring. Not until the parasites and predators have died from lack of food can the aphids multiply unchecked again. Thus very large autumn populations of aphids often follow small spring populations, and small summer populations, large spring ones. Spring populations are usually larger in the south of England than elsewhere, because overwintering is easier; then because of attacks by predators and parasites, summer populations are usually smaller. In the north living aphids overwinter less commonly; spring development is late, and populations develop slowly to reach a maximum in late summer. The largest populations are usually found in the central and east midlands, where the aphids often multiply rapidly before predators and parasites become numerous enough to control them.

Infectivity of aphids bred on infected plants

An attempt was made to determine the proportion of winged aphids, bred on infected cauliflower plants, that could transmit virus when they left the plant. To make the experiment as nearly natural as possible, plants infected with either ClMV or CBRSV were planted outside. The wooden base of a cage was divided into two, so that it could be fitted around the stem of the plant, and the muslin and 'xylonite' upper part of the cage fitted on to this. The cage measured 26 in. × 26 in. × 24 in., and all wooden surfaces were painted white so that aphids could be seen easily (Pl. 6*a*).

The plants were infested with aphids. When winged aphids, bred on a plant, seek to leave it they usually fly towards the brightest light. In the cages they usually flew to the roof or to one side. When aphids were to be tested the cage was cleared of winged aphids, leaving only those on the plant, and then the aphids which flew from the plant were collected at half-hourly intervals. They were placed individually on turnip or Chinese cabbage seedlings under small glass and nylon-gauze covers, and allowed to feed for about 24 hr., after which they and their progeny were killed. The numbers of plants that later showed symptoms are given as Table 18.

TABLE 18. Infections caused by winged aphids transferred to healthy seedlings within 30 min. of leaving infected plants on which they had developed

	No. of aphids	Percentage transmission
Cauliflower mosaic virus		
Myzus persicae		
Young plants 1953	80	16·3
Old plants 1953	111	19·8
Medium plants 1954	87	18·4
Total	278	18·3
Brevicoryne brassicae		
Young plants 1955	83	15·7
Cabbage black ring spot virus		
Myzus persicae		
Medium plants 1954	68	22·1
Old plants 1955	67	4·5
Total	135	13·3
Brevicoryne brassicae		
Old plants 1955	38	2·6

The proportions of ClMV transmissions obtained at different times with *M. persicae* and *B. brassicae* were similar, being between 15 and 20 per cent. The figures suggest that more aphids might be infective when bred on older plants, but more, especially *B. brassicae*, will have to be tested to establish this. Also many more aphids leaving CBRSV-infected plants will have to be tested before it is known if the big differences between the infectivity of those leaving young and old plants are significant. The present limited data, however, suggest that aphids can readily acquire either virus from young infected plants, but that ClMV is more readily acquired than CBRSV from old plants (see page 33).

THE INCIDENCE OF THE DISEASES IN SEEDBEDS

Most winter cauliflower seed is sown during late March and April, and the seedlings are transplanted from late June to early August. Summer cauliflower seed is usually sown in cold frames during autumn, but occasionally in heat during spring. Brussels sprouts seed may be sown in the autumn in sheltered beds, or in frames, or outside in early March. Spring cabbage seed is sown during the second half of July; summer cabbage either in frames in January or in the open during February and March; autumn and winter cabbage during May and June; savoy cabbage mid-March to April. Thus in large market-garden areas there is rarely a period of the year when there are no seedbeds.

Seedlings may become infected in the seedbed and then act as sources

43

of virus within the crop after transplanting. Caldwell & Prentice (1942 b) showed that ClMV causes more loss when plants are infected young than when old, and stressed the importance of keeping seedlings healthy. They found infected plants more often in seedbeds near hedgerows than in the middle of fields, and thought the virus came from hedgerow plants. As they failed to find infected weeds, it may be that hedges, trees and buildings were acting as barriers to aphids, causing more aphids to alight near them.

Control by isolation of seedbeds

Pound (1946) found many cabbage seedlings became infected in beds near old crops; there were still many infections at 1000 yd. distance, but there were few some miles away. Much work on the isolation of seedbeds has been done at the Seale-Hayne Agricultural College, Devon: Jenkinson (1955) compared the yield and the incidents of ClM in plants raised, during successive years, in seedbeds (1) on Dartmoor or the north Devon coast, at least 1 mile from known sources of virus, (2) at least half a mile from brassica crops on the College farm, and (3) adjacent to infected plants on the farm. Many more seedlings were infected in the seedbeds sown near infected crops, but there was no significant difference between the incidence of disease in the seedbeds isolated on the same farm and in those distantly isolated. Plants from the isolated beds yielded much better than those from the beds near infected crops, even if they became infected later.

In Yorkshire during 1951 Dr I. F. Storey and his colleagues in the N.A.A.S. got similar results. A local strain of winter cauliflower was sown in three seedbeds, (1) distantly isolated from brassica crops at High Mowthorpe, (2) in relative isolation at Cawood (other brassicas on the farm), and (3) by the side of a crop at Wakefield in which 80 per cent. of the plants showed ClM. Seven hundred seedlings from each of the three beds were planted in replicated plots at Cawood, and were examined weekly. Plants showing ClM symptoms in July, almost certainly infected in the seedbed, were (1) High Mowthorpe 0 per cent., (2) Cawood 0·3 per cent., (3) Wakefield 5·7 per cent.

Experiments done in Kent have been described by Glasscock and Moreton (1955). The incidence of ClM in cauliflowers raised (*a*) under aphid-proof cages, (*b*) in isolation in Wales, (*c*) in seedbeds surrounded by kale in Thanet, and (*d*) in commercial seedbeds in Thanet, was assessed after transplanting in replicated plots. Plants from under cages and from Wales were healthy when transplanted, and fewer of those from the seedbeds in kale were infected than from the commercial beds;

although ClM incidence increased slowly throughout the growing season in all the plots, these differences were maintained.

In Melbourne, Derbyshire and Woodborough, Nottinghamshire, areas of intensive brassica growing on many small-holdings, between 30 and 70 per cent. of seedlings were often infected with ClMV, whereas in Lincolnshire, in a predominantly agricultural area, infected seedlings were few, from 0·2 to 1·8 per cent. In the small-holding areas ClMV also spreads frequently from crop to crop after transplanting (Mr E. Lester, *in litt.*).

Insecticidal treatment of seedbeds

On small farms it is often impossible to separate brassica seedbeds from standing crops, and, if farmers are to raise their own seedlings, some other control measure than isolation is required. Insecticides are an obvious possibility, but so far have had little success. Caldwell & Prentice (1942*b*) sprayed seedbeds twice with nicotine-soap solution, but failed to decrease the incidence of ClM. Similarly Jenkinson & Glynne Jones (1951) found schradan controlled aphids as pests on seedlings, but it did not affect the incidence of ClM.

Three experiments with insecticides were done at Rothamsted. On 20 June 1951 small seedbeds were sown with cauliflower seed; each consisted of four drills, 6 ft. long and 15 in. apart, and an adjacent area of broadcast seed, 6 ft. long and 2 ft. wide, the same quantity of seed being used on the broadcast and drilled areas. The broadcast seed was raked and rolled in. There were three replicates of four treatments: (1) unsprayed; (2) parathion (*OO*-diethyl *O-p*-nitrophenyl phosphoro-thionate), 0·05 per cent. in aromatic solvent emulsified with 0·1 per cent. Lissapol N; (3) DDT (1:1:1-trichloro-2:2-di(*p*-chlorophenyl) ethane), 0·1 per cent. commercial, formulated in the same way as parathion; and (4) isopestox (bis(mono-*iso*propylamino) fluorophosphine oxide), 0·1 per cent. active ingredient in water containing 0·1 per cent. Lissapol N.

The plots were surrounded, at a distance of three feet, with a row of plants infected with ClMV. All beds were dusted with DDT when the plants were in the cotyledon stage to control flea-beetles. After the first true leaves appeared they were sprayed 'to run-off' every 7 days (i.e. six times), until the plants were large enough to transplant. ClM symptoms usually develop in seedlings about 3 to 4 weeks after infection in midsummer, so the plants were left unsprayed in the seedbeds for another 4 weeks, when they were lifted and examined. Although some of the sprays decreased the number of infected plants, these effects are not significant at the 5 per cent. level (Table 19).

TABLE 19. Percentages of plants infected with ClMV and CBRSV in seedbeds treated with insecticides, 1951 and 1952

Treatment	1951	1952 (1)		1952 (2)
	ClMV	ClMV	CBRSV	ClMV
Nil	14·3	1·1	2·0	56·7
Medium alone	.	2·5	2·8	48·6
DDT	9·6	.	.	50·5
Toxaphene	.	1·3	1·5	.
Pyrethrins	.	1·3	1·2	55·4
Parathion	8·0	1·7	1·2	.
Pyrolan	.	0·7	0·5	43·8
Isopestox	10·3	.	.	.
Schradan	.	2·1	1·2	.
Systox	.	1·8	1·2	60·7

In 1952 the first experiment was sown at the beginning of May, the second in mid-August. In the first, eight treatments were applied to seedbeds consisting of three 3-ft.-long rows, 1 ft. apart; each treatment was replicated seven times. The experiment was suitably randomized into blocks, which were surrounded by single rows of cauliflower plants, 5 ft. from the beds, infected with ClMV and CBRSV alternately. The beds were sprayed 'to run-off' four times, at 9- or 10-day intervals, the treatments being: (1) unsprayed; (2) the medium, 0·5 per cent. aromatic solvent (equal parts of benzene and toluene) emulsified with 0·1 per cent. Lissapol N; (3) toxaphene (chlorinated camphene), 0·125 per cent. in the medium; (4) pyrethrins, 0·1 per cent. piperonyl butoxide, 0·03 per cent. in the medium; (5) parathion 0·05 per cent. in the medium; (6) pyrolan (3-methyl-1-phenyl-pyrazolyl-(5)-dimethylcarbamate), 0·1 per cent. in the medium; (7) schradan (*bis*(dimethylamino)phosphonous anhydride), 0·1 per cent. in water emulsified with 0·1 per cent. Lissapol N; and (8) systox (*O,O*-diethyl *O*-ethylmercaptoethyl phosphorothionate) 0·1 per cent. active ingredient (own wetter included).

Infections were few in the first experiment in 1952 (Table 19); no treatment produced any significant result, but there was a tendency for more plants to be infected when sprayed with the medium alone, and for fewer among those sprayed with pyrolan.

In the second experiment the seedbeds were doubled in size to three rows, 6 ft. long, and there were six replications of six treatments, formulated as in the previous experiment: (1) unsprayed; (2) the medium alone; (3) DDT, 0·2 per cent.; (4) pyrethrins, 0·03 per cent. +piperonyl butoxide 0·03 per cent.; (5) pyrolan, 0·1 per cent.; and (6) systox, 0·1 per cent. The beds were sprayed 'to run-off' four times, at 10–14-day intervals. Many plants contracted both ClMV and CBRSV in this experiment, but as recognition of 'mixed' symptoms was made difficult by *B. brassicae* feeding damage on the untreated plants, the

46

recording was confined to ClM (Table 19). Pyrolan again decreased infections more than any other spray, but none of the differences was significant.

In these experiments there was little spread of virus within the seed-beds, even in the unsprayed beds when aphids were very numerous. It is concluded that, apart from limiting the damage caused by infestations of *B. brassicae*, spraying seedbeds with either contact or systemic insecticides has no practical value.

Spacing and size of seedlings

In the insecticide experiments the seedlings were graded as large, medium, or small, the medium being suitable for transplanting, and the others too big or too small. The number of plants in each grade

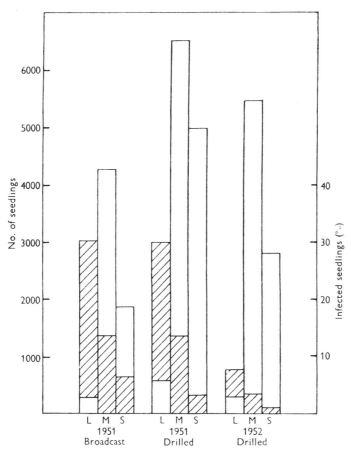

FIG. 12. Numbers of seedlings ☐, and percentages of infected seedlings ▨, in the Rothamsted insecticide experiments, L=large, M=medium, S=small seedlings.

depended upon the density of sowing. In 1951, when the same quantity of seed was sown broadcast and drilled, almost twice as many plants developed in the drilled as in the broadcast plots (Fig. 12), but a greater proportion (though not a greater number) of the broadcast plants was suitable for transplanting. The seed-rate for drilling (42 seedlings per foot of row) was too high. The percentages of infected plants in the broadcast and drilled plots were similar; more large plants were infected than medium-sized, and fewer small plants. Similar proportions were obtained in the first experiment in 1952 (Fig. 12). The seed-rate used in this experiment was lower (17 seedlings per foot of row), and more plants were suitable for transplanting.

In 1952 an experiment was sown at the end of April to test the effects of seed-rate and row-spacing on infection. The treatments, applied to small seedbeds $4\frac{1}{2} \times 3$ ft., were:

(1) Light seed rate   ×6 rows spaced 9 in. apart.
(2) Medium seed rate ×6 rows spaced 9 in. apart.
(3) Heavy seed rate   ×6 rows spaced 9 in. apart.
(4) Light seed rate   ×3 rows spaced 18 in. apart.
(5) Medium seed rate ×3 rows spaced 18 in. apart.
(6) Heavy seed rate   ×3 rows spaced 18 in. apart.
(7) Light seed rate   ×broadcast (allowing $1\frac{1}{2}$ times the weight of seed as 1).
(8) Medium seed rate ×broadcast (allowing $1\frac{1}{2}$ times the weight of seed as 2).
(9) Heavy seed rate   ×broadcast (allowing $1\frac{1}{2}$ times the weight of seed as 3).

There were four blocks of the nine treatments randomized, and each block was surrounded at a distance of 5 ft. by plants infected with ClMV and CBRSV alternately. The average numbers of seedlings that developed per foot of row with the different seed-rates were: light 10, medium 16, and heavy 28. The numbers and percentages of large, medium and small seedlings in the different treatments are shown in Fig. 13. Although in the broadcast plots 150 per cent. as much seed was sown as in the 9-in.-spaced drills, only 56 per cent. as many seedlings developed. The percentages of seedlings suitable for transplanting fell rapidly as the seed-rate was increased. There was little difference in the incidence of the two diseases and they have been added together to give the percentages of plants infected in Fig. 14. As the plant population increased, so the disease incidence fell. The plots with rows 18 in. apart and the light seed-rate contained the highest proportion of infected plants; the least was with the heavy seed-rate at 9 in. The incidence of

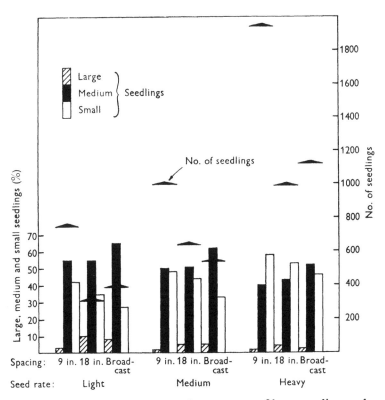

FIG. 13. Numbers of seedlings, and percentages of large, medium and small seedlings at different seed rates and spacing.

disease was too low for these results to be significant, but they agree with the results of van der Plank (1947) and Blencowe & Tinsley (1951).

The first insecticide trial in 1952 was so designed that one outer row of each plot faced the row of diseased plants five feet away, whereas the other outer row faced that of an adjacent plot two feet away. More plants in the outer rows near the diseased plants were infected than in the other two rows in all treatments, the mean percentage of infected plants in all outer rows being 4·6, in the other rows 1·9. More plants were infected in the outer rows than in the inner rows of unprotected seedbeds in the barrier trials (see below).

The following conclusions can be drawn from these experiments: (1) more plants are obtained per pound of seed by drilling than by broadcasting; (2) the percentage of plants infected (but not the total number) will be lower if drills are put closer together; (3) increasing the seed-rate decreases the percentage of infected plants, and also decreases the percentage of seedlings suitable for transplanting; 15 to 20 plants per foot of row appears to be about optimal; (4) the largest plants are the most likely to be infected, and should be rejected when transplanting; and (5) more plants are infected in outer rows of seedbeds than in inner rows, and outside rows should not be transplanted if there is a surplus of seedlings.

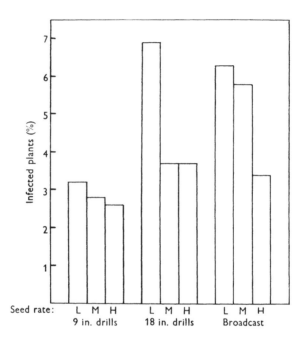

FIG. 14. Percentages of infected seedlings at different seed rates (L=light, M=medium, H=heavy) and spacing.

Control by barrier crops

That a cover crop will protect an undersown one against insect-borne viruses has long been recognized, but the idea of using plant barriers to protect seedbeds arose from recent epidemiological studies on virus diseases of potato, sugar beet, lettuce and other crops. The aphids spreading viruses from one crop to another often land on the outermost rows of a crop, so that a higher proportion of the edge-row than central plants become infected. This suggested that, if a brassica seedbed were surrounded by other, preferably taller, plants, arriving aphids might react in one of the following ways, illustrated in Fig. 15. They might land on the barrier plants and either stay there (*a*), or fly away (*b*). Aphids usually probe a plant immediately they alight, and in doing so might free themselves from a non-persistent virus, and then be unable to infect plants in the seedbed to which they might move later (*c*). Aphids often fly upwards when leaving a plant (*d*), and many will fly away over the seedbed when they leave the barrier. Thus, the barrier could protect the seedbed from viruses both by decreasing the total number of aphids that alight on the seedlings and by decreasing the number of infective aphids that alight.

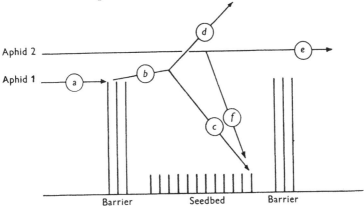

FIG. 15. Diagram illustrating the influence of plant barriers around seedbeds on the movements of aphids (see text for explanation).

If the barrier plants are suitable food plants for the aphids, more would stay than if not, and more would feed longer even if they eventually left, so that a greater proportion of the aphids leaving would be non-infective. If, however, the plants are susceptible to the same virus as the seedlings, they could become infected and then act later as sources of virus.

Taller barrier plants will cause wind-eddies, and aphids passing over them may be blown straight over the seedbed (e), or may be deposited on the seedlings (f), but experiments on snow-fences (Jensen, 1954) show that a barrier gives substantial protection for a distance at least ten times its height.

Brassica seedbeds are often drilled in long narrow strips which lend themselves to protection by barriers. The drilling of seedbeds within another crop (e.g. the beds placed in kale crops in Kent in 1950, page 44) is one way of making use of the barrier effect. This is not always practicable, however, especially on smallholdings, but most farmers could easily drill a few rows of barrier plants around their seedbeds. The following experiments were done, therefore, in different parts of England to determine the effectiveness of such barriers in reducing the incidence of disease in the seedbeds.

Small cauliflower seedbeds, four rows 6 ft. long and 12 in. apart, were either unprotected or surrounded by one of the following barriers :

(1) 3 rows of mustard, 8 in. apart ⎱ the inner row being one foot from
(2) 4 rows of barley, 6 in. apart ⎰   the outer seedbed row.
(3) Either 'Sisalkraft' tarred paper, ⎫
   32 in. high, or hop-lewing ⎬ 30 in. from the outer seedbed row.
   (coarse string netting) hung
   double, 28 in. high ⎭

The treatments were replicated seven times (four Sisalkraft + three hop-lewing), the seedbeds being sown in blocks of four (unprotected, mustard, barley and 'physical'). Rows of cauliflowers were planted between each block, 5 ft. from the barriers, alternate plants being infected with either ClMV or CBRSV.

The barrier crops were sown on 24 April, the cauliflower seed on 15 May. The mustard eventually seeded and reached a height of about 52 in., the barley a height of about 27 in.

The seedlings were examined four weeks after they were ready for transplanting. *B. brassicae* damage was common on the unprotected seedlings and on those surrounded by 'physical' barriers, but not on those surrounded by barley and mustard, indicating that the plant barriers had decreased the infestations.

The mean percentage infection with viruses in the different plots is given in Table 20. Some plants were infected with both viruses, and the percentage of infected plants is also given. The percentages in the beds surrounded by Sisalkraft and hop-lewing were similar and are recorded under 'physical' barriers.

TABLE 20. Percentage of infected plants in the Rothamsted seedbed barrier experiment No. I, 1952

Barrier	Cauliflower mosaic	Cabbage black ring spot	Infected plants
None	4·5	5·2	8·6
Mustard	0·7	21·3	21·6
Barley	0·4	0·6	0·9
'Physical'	2·2	3·4	5·0

The barley and mustard were sown too near the cauliflower, which made very poor growth in the outer rows; the seedbeds were small in relation to the size of the barriers, but the results show that narrow barriers can significantly reduce the incidence of virus diseases, and that plant barriers are much more effective than physical ones. The mustard was heavily infested with *M. persicae*, and was infected with both viruses, but especially CBRSV. As it matured its leaves withered and fell, and presumably the aphids walked on to the cauliflower seedlings and transmitted CBRSV to them.

Again more of the larger than of the smaller cauliflower seedlings became infected; in the unprotected seedbeds percentage infection was: ClM, large plants 37·5, medium 6·5, small 0·5; CBRSV, large plants 13·8, medium 7·0, small 2·4.

As soon as the first experiment was finished another was laid out on the same ground to find out if treating mustard with a systemic insecticide (schradan) would make it a more effective barrier. The seedbeds were the same size as before, four rows 6 ft. long, and were either unprotected or surrounded with three rows of mustard, 8 in. apart. Half the mustard barriers were sprayed at intervals with schradan. The inner barrier rows were 30 in. from the outer seedbed rows.

The mustard was sown on 12 August, the cauliflower the next day. By 12 September the mustard was 8 in. high, but it was infected, especially with CBRSV, and by the end of September many of the plants were dead in both the sprayed and unsprayed barriers. The incidence of disease in the cauliflower seedlings on 10 October is given in Fig. 16. Many plants in the unprotected beds were damaged by *B. brassicae*, but few of those surrounded by

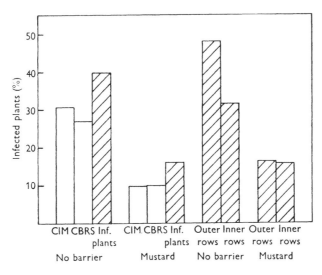

FIG. 16. Percentages of infected plants in the Rothamsted seedbed barrier experiment, No. 2, 1952.

mustard. The late-sown mustard failed to mature, and apterous aphids did not move and spread virus from the barriers, so there was little difference in disease incidence between the sprayed and unsprayed mustard plots. The outer rows of the unprotected seedbeds acted as barriers for the inner rows in both experiments; differences between outer and inner rows were small in protected beds.

CAWOOD BARRIER EXPERIMENT, 1952

Messrs F. G. Smith and J. P. Cleary tested barriers on the Stockbridge House Experimental Horticulture Station, Cawood, Yorks. The experimental area was in the middle of a field of spring wheat. There were six long plots, 6 yd. × 104 yd., separated from each other by strips of wheat 9 ft. wide. Three rows of cauliflowers infected with ClMV were planted across the centre of each plot, dividing it into two sub-plots; four-row cauliflower seedbeds were sown across each sub-plot on 22 May, at a distance of either 20 yd. or 40 yd. from the infected plants, and the intervening area, and the area beyond, were left bare or were sown with either wheat (on 28 February) or kale (on 1 March).

The seedlings were ready for transplanting by 3 July; they were left until 20 August, when the plants in the two rows nearest to the infected plants were examined (Fig. 17).

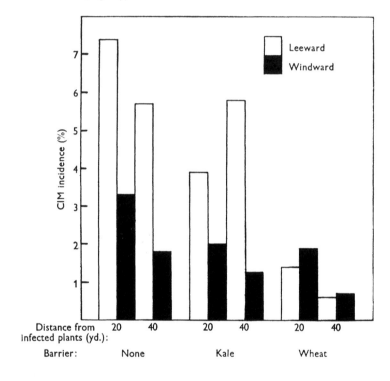

FIG. 17. Incidence of ClM in the Cawood seedbed barrier experiment, 1952.

Sub-plots were either to windward or leeward of the infected plants, and seedbeds to leeward had more infected plants, except within the wheat barriers. As at Rothamsted, the non-susceptible cereal barrier protected the seedbeds better than the susceptible brassica barrier. Disease incidence was less at 40 yd. than at 20 yd. from the virus source, except when the seedbeds were in kale, which itself became infected.

ROTHAMSTED BARRIER EXPERIMENT, 1953

Seedbeds surrounded by narrow barriers (3 rows, 1 ft. apart) of barley, sown before, and at the same time as the cauliflower seed, were compared with unprotected beds. There were four replicates of each treatment, two on either side of a central double strip of plants infected with ClMV, 10 ft.

from the nearest barriers. The plots consisted of (a) a 16-row bed, with 10 ft. rows, 1 ft. apart, and (b) two 5-row beds with similar rows. In the protected plots each bed was surrounded by a barrier, and the two 5-row beds were adjacent to, and occupied the same area as, the 16-row bed. All the barriers were 30 in. from the nearest cauliflower row.

The early sown barley (which reached an average height of 34 in.) was drilled on 21 April, the late sown barley (which reached an average height of 19 in.) and the cauliflower seed during the second week of May. Aphids were few during the summer, and few plants were infected. Every seedling in the middle 4 ft. of each row was examined during 18–20 August (Table 21). The

TABLE 21. Percentage incidence of ClM in the Rothamsted seedbed barrier experiment, 1953

Barrier	None	Barley	
		Pre-sown	Late-sown
		All seedlings	
5-row seedbeds	2·1	0·4	0·4
16-row seedbeds	1·2	0·3	0·2
Mean	1·6	0·4	0·3
		Seedlings suitable for transplanting	
ClM incidence	2·1	0·5	0·5
Percentage plants usable	62	67	68

figures are too small to be significant, but there were four times as many infected plants in the unprotected beds as in those with barley barriers. The pre-sown, taller barley was no better than the late-sown barley, and the 16-row seedbeds were protected as adequately as the 5-row beds. More usable plants were obtained from the protected beds, probably because the barley protected the seedlings and soil against cold, drying winds.

To find out if the barrier effect of the barley was purely physical, or if aphids fed and were therefore likely to lose their infectivity, winged *M. persicae* and *B. brassicae* were placed on barley plants and were observed to probe, and presumably feed, for periods varying from 20 sec. to 28 min. Apterous *M. persicae* occasionally gave birth to young on the barley, but these did not survive many days.

OTHER SEEDBED EXPERIMENTS, 1953

Experiments with the same design were done by Dr H. E. Croxall near Newcastle, by Messrs F. G. Smith and J. P. Cleary at Cawood, by Messrs H. Fairbank, N. Moss and the writer at Luddington, by Messrs P. H. Brown, G. F. Wheeler, W. Buddin and the writer at Lymington, by Dr H. H. Glasscock and Miss U. E. K. Fraenkel at Richborough, and by Mr J. G. Jenkinson at Seale-Hayne Agricultural College.

The design was basically similar to that at Rothamsted, seedbeds of six rows, 8 ft. long, being drilled on either side of a double strip of plants infected with ClMV. There were two replicates of each of the following barrier treatments at each place, i.e. twelve replicates in all : (1) no barrier; (2) marrowstem kale; (3) marrowstem kale treated with a systemic insecticide; (4) barley and (5) broad beans. Each barrier consisted of three rows, 8 in. apart. The kale was sown in late March, the barley and beans in early March, and the cauliflower seed in mid-April. There were spaces of 10 ft. between the inner barriers and the infected plants, of 7 ft. between the side barriers of each seedbed, and 30 in. between the barriers and the seedbeds.

Aphid traps were placed over bare ground near the centre of each experiment (Table 22).

TABLE 22. Aphids caught on flat sticky traps in seedbed experiments 1953, from late April or early May until four weeks before recording

	M. persicae	B. brassicae	Others
Newcastle	0	0	119
Cawood	0	1	18
Luddington	0	0	114
Richborough	7	0	235
Lymington	1	1	169
Newton Abbot	3	11	1134

At least four weeks after the plants were ready for transplanting all the seedlings in the middle four feet of each row were examined. The incidence of ClM is given in Table 23. M. persicae and/or B. brassicae were fairly numerous only at Richborough and Newton Abbot, and only at these places did virus spread sufficiently to give results.

TABLE 23. Percentage incidence of ClM in six seedbed experiments with the same design in different parts of England, 1953

	Barrier				
Site	None	Kale	Kale + insecticide	Barley	Beans
Newcastle	0	0	0	0	0
Cawood	0·7	0	0	0	0·1
Luddington	0·2	0	0	0	0
Richborough	7·0	3·1	3·9	1·1	2·2
Lymington	0·1	0	0	0	0
Newton Abbot	9·7	1·9	1·5	2·4	3·8
Mean	3·0	0·8	0·9	0·6	1·0

The barriers were again successful in partially protecting the seedlings from infection. Treating kale with a systemic insecticide did not increase its efficiency as a barrier. On average, barley was the most efficient barrier, reducing infection to one-fifth of that in the unprotected beds.

An unexpected result in the Rothamsted experiment was the higher proportion of usable seedlings in the beds protected by barriers. Similar results were obtained at Newcastle, Cawood and Lymington. Table 24 gives the percentages of usable plants at the four places where this information was obtained.

TABLE 24. Percentages of usable plants in the 1953 seedbed barrier experiments

	Barrier			
	None	Kale	Barley	Beans
Newcastle	69	76	88	77
Cawood	51	64	65	64
Lymington	56	64	65	64
Luddington	64	67	64	64
Mean	60	68	71	67

In most of the previous experiments a cereal proved the best barrier, whether the seedbed was placed in a crop, or was surrounded by three barrier rows. During 1954 an attempt was made to test this method of control in a randomized experiment, using a technique that might be applied in commercial practice.

The common six-spout drill was used with either (1) the two outer boxes delivering barley seed and the four inner boxes cauliflower seed, to give four 4-row seedbeds per plot, protected on the outside by single rows of barley and divided from each other by two rows of barley, or (2) 22-row beds protected by single rows of barley. The individual plots were 25 ft. long and 23 ft. wide, and were drilled in four blocks of contiguous plots, two blocks of 4-row beds and two of 22-row beds. Each block contained two replicates of the following treatments, grouped in sub-blocks of three plots, all suitably randomized: (a) early-sown barley, sown six weeks before the cauliflower seed, (b) late-sown barley, sown at the same time as the cauliflower, and (c) no barrier. A single drill of barley was also sown across the seedbed drills, separating adjoining plots, to reduce lateral spread of virus inside the barriers. Rows of plants infected with ClMV were planted on each side of each block, 8 ft. from the barriers. The cauliflower seed was not sown until 25 June, so that the seedlings should be growing when aphids might be expected to be most plentiful.

The incidence of ClM was recorded in two 3-ft.-wide strips of seedlings across alternate drills of each seedbed about five weeks after the seedlings were ready to transplant (Fig. 18). Only two of the small unusable seedlings were infected, and incidence is given as the percentage of usable seedlings. Aphids were few; during the eight weeks when seedlings were growing, of 391 aphids trapped only two were *M. persicae* and none *B. brassicae*. The weather was cool and wet; the cauliflower seedlings grew well and the beneficial effect of barriers shown in 1953 did not recur.

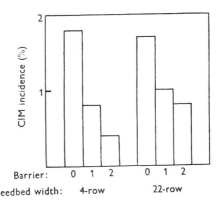

FIG. 18. Incidence of ClM in usable seedlings in the Luddington seedbed barrier experiment, 1954. (0=no barrier, 1=late-sown barley, 2=early-sown barley.)

The differences between the incidence of ClM in the unprotected and protected seedbeds are significant (5 per cent. level) for both the 4-row and 22-row beds, but not the differences between the early-sown and late-sown barrier plots.

An identical experiment was done at Lymington, but infected seedlings were so few that no figures are given.

If narrow barriers of cereal can reduce disease incidence to about one-fifth of that in unprotected seedlings, and in some years give a higher proportion of usable seedlings, they are well worth using commercially, because cauliflower plants infected in the seedbed rarely produce marketable curds, and as is shown below, the ultimate disease incidence in the transplanted crop is often proportional to the number of infected seedlings transplanted.

Cage protection of seedlings

Rigid-framed cages, covered with 28-mesh 'Tygan' nylon screencloth, which excludes aphids, were used at Rothamsted and Luddington to protect seedbeds. So long as the seed was not sown too thickly, the seedlings grew well and were not unduly drawn. The cages were removed a few days before the plants were transplanted, to allow the seedlings to harden off. Mr P. H. Brown, at the Efford Experimental Horticulture Station, successfully modified the idea for a larger seedbed by erecting a triangular structure of Dutch light frames and securing a strip of the nylon screencloth down each side with drawing-pins.

The cages were expensive, but have lasted for five seasons. Few farmers would consider the outlay justified, however, especially when healthy seedlings can be raised by the use of barrier crops and some measure of isolation.

VIRUS SPREAD INTO THE CROP

Not only is virus carried from crops to seedbeds, but often also from one crop to another. In the West Riding of Yorkshire, where cauliflowers are grown in small areas on many small farms, winter cauliflowers mature from mid-May to mid-June, and most of the farmers keep a few plants for seed; seedbeds are sown during April, so for about five months of the summer plants of the old and young crops grow near together. Storey & Godwin (1953) found that most of the plants in young crops that became infected were in the outside rows, nearest to the old crops. The steep gradient of infection within the crop was little affected by the size of the infected 'source' crop or by its distance from the young crop, at least up to fifty yards over bare ground.

Such gradients are typical for the spread of virus by aphids, especially non-persistent viruses, because aphids flying low from one crop to another alight and infect the outer rows, and soon lose their infectivity. This type of spread was lessened with lettuce mosaic virus, by planting crops of similar age in large blocks, separated from older plantings by as great a distance as possible (Broadbent, Tinsley, Buddin & Roberts, 1951). The effect of taking similar measures with brassica crops was studied on a farm at Datchet, in the Thames Valley, during the seasons 1950–53.

During the 1950–51 season the distribution of crops on the farm was typical of that on most horticultural holdings, the crops of autumn and winter cauliflowers, purple sprouting broccoli and cabbages being planted wherever crop rotation and space allowed. In October the incidence of ClM in various brassica crops varied from 9 to 25 per cent. Wherever the crop was, or had been, adjacent to an older crop, ClM was commonest in the rows nearest to that crop (Fig. 19). Infection

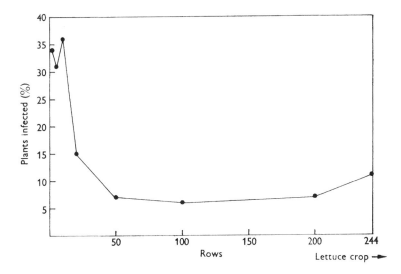

FIG. 19. Incidence of ClM in different parts of a spring cabbage crop adjacent to an old infected winter cauliflower crop, Datchet, May 1951.

gradients of this type are not usually found when spread is from sources within the crop, unless aphids are more numerous on one side of the crop than on the other. Untransplanted broccoli seedlings had been left to grow in the seedbeds, and 73 per cent. of these showed mosaic symptoms in early November.

At the end of February neither the previous year's seedbeds, nor autumn cauliflower and cabbage crops, had been cleared. The old broccoli seedbeds were not destroyed until mid-June, when every plant was infected. In mid-April the incidence of ClM in winter cauliflower crops varied from 17 to 43 per cent. Summer cauliflowers were planted near the old seedbed, and some plants near this became infected. As aphids were few during the spring of 1951, few plants in the young crops became infected, despite the high incidence of ClM in the old crops.

The farmer was persuaded to plant all his winter cauliflower seedlings in one large block in 1951. The seedbed, however, was near the summer cauliflower, and some seedlings became infected, with the result that virus spread from these when aphids became numerous during the autumn.

Surveys in the main English winter-cauliflower growing areas, during the seasons 1950–2, showed that this picture of virus spread from crop to crop, and from crop to seedbed, followed by spread within the crop, was found in all typical horticultural areas, except Cornwall. ClM was uncommon in Cornwall because most winter cauliflowers are cut from November to February, and few other brassicas are grown, even in allotments.

Brassica virus diseases were particularly prevalent in areas where brassica seed was grown. These plants are in the ground for a long period and are often infected. *B. brassicae* infest the flower stems (Markkula, 1953), from which they can acquire ClMV. Not only are infected seed crops a source of virus for other crops in the area, but infected plants yield little seed. Jenkinson (1955) found that healthy Roscoff cauliflowers yielded an average of 16 gm. of seed per plant, but that plants infected with ClMV before October yielded an average of only 0·7 gm. per plant.

Not all virus-carrying aphids land on crops adjacent to those they leave, and virus is sometimes carried long distances; but the aphids are dispersed by wind, and few usually reach fields isolated from other brassicas. Sometimes, however, very large populations of *B. brassicae* (and occasionally of *M. persicae*) develop on brassica crops during July, August and September, and then, if crops are infected, virus may be carried over long distances from crop to crop. The very serious outbreaks of ClM during the period 1948–50 occurred during years when *B. brassicae* were very numerous; overwintering aphids and eggs were particularly numerous during the 1947–8 winter (Empson, 1952).

Although some cruciferous weeds and ornamental plants are susceptible to ClMV, no evidence has been obtained that they play any significant part in the infection story. Pound (1946) reached a similar conclusion in the U.S.A. CBRSV has a much wider host range, including perennial ornamental plants, which undoubtedly provide sources of infection. At Datchet, ClM was widespread during 1952, but CBRS was common only near the farmhouse garden. The widespread occurrence of this virus is shown by the frequency with which wallflower blossoms 'break'. Seed from diseased wallflowers was sown, but no evidence of seed-transmission of the virus was obtained, and it is assumed that this virus is widely aphid-transmitted, especially in urban areas, but only exceptionally does a serious outbreak occur in agricultural areas.

Caldwell & Prentice (1942b) studied virus spread in Devon, and found that it occurred from early infected plants throughout the late summer and autumn. They recommended early roguing to remove the initial sources of infection. Storey & Godwin (1953) contrasted the pattern of infection in fields where there were no adjacent sources of virus, and where the initial sources were probably transplanted from seedbeds, with that in fields which had sources nearby. Instead of the steep gradient of infection from the outside rows towards the centre of the crop, found when there are nearby sources of virus, these fields had infected plants distributed fairly evenly throughout, with a tendency for the incidence of ClM to be lowest in the outer rows, and highest in the middle.

The spread of virus from infected plants within the crop was studied in commercial fields and in specially planted experiments during the years 1951–3. The following numerical key was used in the field studies. It was sometimes difficult to assign a plant to a particular category, but generally the key worked well and the figures obtained gave some indication of the stage of development of symptoms, and of the degree of tolerance to infection.

CAULIFLOWER MOSAIC SYMPTOM KEY

(0) No symptoms.
(1) Definite virus symptoms, e.g. vein-clearing, on one or two leaves.
(2) Symptoms on many leaves, but no stunting.
(3) As 2, but plants somewhat smaller than average.
(4) As 2, but with severe stunting or leaf drop.
(5) Plants dead.

Time and pattern of virus spread

Virus is spread within the crop at any time after transplanting when aphid vectors are active. In England there are usually flights of many aphid species from winter to summer hosts during May and June, and virus is carried into seedbeds and from crop to crop. The major flights of the year are from one summer host to another during July and August, and virus spread is then common both from crop to crop and within crops. Spread in winter cauliflower crops at this time is particularly important, because plants infected when young yield little, and act as sources of virus for later spread. Winged aphids are active through September and October, when many species are flying to winter hosts, but because plants then grow slowly, symptoms often fail to show until spring, and yield is not greatly affected. There is considerable variation from year to year, and from one part of Britain to another, in this generalized picture of events, as is shown in Tables 25 and 26.

TABLE 25. Aphids trapped per month on cylindrical sticky traps in experiments on the spread of ClMV

(M.p. = M. persicae; B.b. = B. brassicae.)

County...	Buckingham			Devon			Hampshire			Hertford			Kent			Warwicks			Yorks		
	M.p.	B.b.	Other	M.p.	B.b.	Other	M.p.	B.b.	Other	M.p.	B.b.	Other	M.p.	B.b.	Other	M.p.	B.b.	Other	M.p.	B.b.	Other
1951																					
July	12	4	1127	22	1	396	.	.	.	5	0	1425	14	0	755	8	0	978	3	0	729
August	7	50	272	2	17	204	.	.	.	5	2	389	3	0	241	5	3	460	8	0	880
September	0	1	29	2	0	54	.	.	.	3	1	92	6	0	148	3	45	87	10	0	745
October	2	0	16	1	2	61	.	.	.	4	6	39	35	0	120	10	12	68	2	0	38
November	0	0	17	3	0	10	.	.	.	0	0	5	2	0	1	0	0	12	2	0	7
1952																					
June	1	0	312	.	.	.	.	.	.	5	1	877	.	.	.	.	.	.	4	0	672
July	2	11	407	7	31	67	4	2	108	4	64	793	2	17	101	7	403	186	9	21	1678
August	3	76	91	6	65	59	2	1	130	23	349	58	3	21	48	2	18	19	3	25	130
September	3	20	96	3	1	29	2	2	34	1	13	39	0	1	23	15	0	46	21	84	125
October	2	0	56	8	3	47	3	0	15	.	.	.	0	0	2	0	0	1	1	5	31
November	0	0	0	3	0	19	.	.	.	.	.	.	0	0	2	.	.	.	.	.	.

62

TABLE 26. Percentage increase in ClM incidence in winter cauliflower crops

County	Buckingham		Kent	Warwicks	Devon	Yorks
Variety	Mid-summer	Con-tinuity	White Beauty	St George	Roscoff A6	Local strain
1951–2						
July	.	.	.	.	0·5	0·5
August	.	.	0·6	0·6	.	1·1
September	1·6	1·3	0·8	.	1·2	4·3
October	6·3	3·6	1·1	1·0	1·6	11·5
November	12·2	6·4	1·2	2·8	1·8	.
December	.	.	1·3	4·9	2·0	.
January	.	.	1·3	.	2·1	.
February	.	.	1·4	.	.	.
March	34·5	15·5	1·7	17·3	.	.
April	.	.	.	.	.	.
May	.	.	.	.	.	19·0

County	Buckingham			Hamp-shire	Kent	Warwicks	Devon
Variety	Con-tinuity	Mid-summer	Reading Giant	Extra Early Roscoff	Satis-faction	Con-tinuity	Roscoff A6
1952–3							
July	.	.	.	0·07	.	.	0·5
August	.	.	.	.	0·1	0·1	.
Sept.	2·3	4·9	5·7	0·4	0·2	40·0	2·1
Oct.	8·4	14·7	14·0	1·0	0·6	.	5·4
Nov.	14·5	.	.	4·3	1·5	.	8·2
Dec.	.	.	.	.	1·9	.	9·6
Jan.	.	.	.	.	2·1	87·0	10·2
Feb.	.	.	.	.	2·5	.	.
March	.	.	26·4	.	4·0	.	.
April	22·2	39·6	.	.	.	99·0	.

Early-planted crops sometimes have more infected plants than late-planted crops because they are in the field when the summer aphids migrate; for instance, a crop transplanted in mid-July with about 0·4 per cent. of the seedlings infected, had 4 per cent. of the plants infected in late September, whereas a crop transplanted from the same seedbed in early August had 0·8 per cent. only.

FIELD EXPERIMENTS IN 1951–2

Experiments on the spread of ClMV were done in Buckinghamshire, Devon (Jenkinson, 1955), Kent, Warwickshire and Yorkshire.

Buckinghamshire. The incidence of ClM was recorded at intervals in winter cauliflowers (varieties Continuity, Midsummer and Reading Giant), which were planted in one large block at Datchet. All the plants in five single rows through the crop were examined, in addition to plants in three areas (15×15 plants) which were marked out around seedbed-infected plants. Winged *B. brassicae* were numerous during the first and last weeks of August (Table 25), but they had developed on summer crops, few of which were infected, so that winged aphids had little opportunity to spread virus to the winter cauliflower crops. However, a few seedlings were infected in the seedbed, and the aphids flying within the crop soon spread virus from these, especially to adjacent plants (Fig. 20).

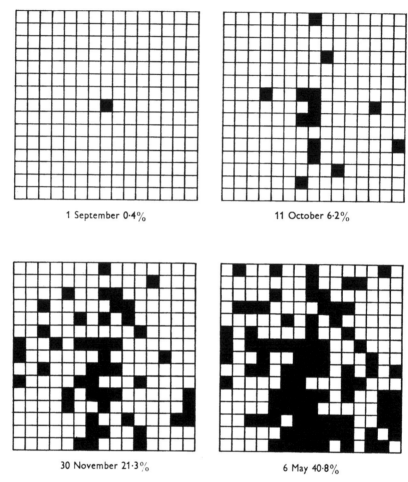

1 September 0·4%   11 October 6·2%

30 November 21·3%   6 May 40·8%

FIG. 20. Spread of ClMV in a winter cauliflower crop, variety Midsummer, Datchet, 1951–2.

The variety Midsummer was much more susceptible than Continuity, as
aphids were equally numerous on both varieties, and the incidence of ClM
was initially about 1 per cent. in both (Table 27). Reading Giant, transplanted
just after the main aphid flights, had fewer infections.

Recording was complicated by 'tip-burn' symptoms that developed during
the late autumn, and rendered the plants liable to heart rot. Midsummer
suffered more than Continuity, both virus-infected and uninfected plants
being killed. Some of the dead plants (symptom category 5) in Table 27 had
been killed by this, but the figures are sufficiently striking to show that
Continuity tolerates virus infection better than Midsummer. The date of
infection had little effect on the final symptoms in Continuity, and about
25 per cent. of the infected plants were total failures (symptom categories 4
and 5), whereas in Midsummer 96 per cent. of early-infected plants were
total failures, compared with 48 per cent. of late-infected plants.

64

TABLE 27. Development of ClM symptoms in two varieties of winter cauliflower,
Datchet, 1951–2

Variety	Date symptoms first seen, 1951	Percentages of infected plants in spring with symptom nos. (see key in text)			
		2	3	4	5
Continuity (15·5 per cent. ClM on 18 March)	6–14 Sept.	17	58	25	0
	11 Oct.	15	54	23	8
	9 Nov.	21	54	25	0
Midsummer (34·5 per cent. ClM on 18 March)	6–14 Sept.	0	4	32	64
	11 Oct.	0	47	33	20
	9 Nov.	0	52	31	17

Kent, Warwickshire and Yorkshire. Similar experiments were done by N.A.A.S. officers in the Isle of Thanet, Kent, at Luddington, Warwicks., and at Cawood, Yorks. There were five blocks of five treatments, the plots being 10 or 11 plants square. Treatments were:

(1) No infected plant introduced.
(2) A plant of the same age as the crop, infected with ClMV, was planted at the same time as the crop, near the centre of the plot.
(3) As 2, but the 'infector' was removed in October.
(4) As 2, but the 'infector' was not planted until October.
(5) As 2, but the 'infector' was an old plant when transplanted.

In Kent *M. persicae* were numerous during October, but they spread little virus (0·6 per cent. ClM at planting increased to 1·7 per cent. on 11 March 1952); nearly half of them were males, and it is probable that those trapped were leaving the crop, and flew little from plant to plant. No *B. brassicae* was trapped.

At Luddington the time when symptoms appeared corresponded well with the aphid trap catches (Tables 25 and 26). The general incidence of ClM increased from 0·6 per cent. at planting to 17·4 per cent. on 12 March 1952. At Cawood ClM increased from 0·5 to 19·0 per cent. on 22 May 1952. No *B. brassicae* was trapped in this experiment, and the virus was spread by *M. persicae* which were active throughout the summer and autumn (Table 25).

In the three experiments there was hardly any spread of virus after the 'infectors' in treatments (3) and (4) had been changed, i.e. after mid-October, and there was little difference in the effectiveness as virus sources of the young and old 'infectors'. At Luddington there was not much reduction in curd size in plants of the variety St George infected with ClMV, except when plants were infected in the seedbed or soon after transplanting.

FIELD EXPERIMENTS IN 1952–3

Experiments on virus spread were done in Buckinghamshire, Devon (Jenkinson, 1955), Hampshire, Hertfordshire, Kent and Warwickshire.

Buckinghamshire. An area of winter cauliflower was 'rogued', i.e. all diseased plants were pulled and carried from the field in sacks. The area ran across three strips of cauliflower, 31 rows of Reading Giant, planted in early August, 24 rows of Continuity, planted mid-August, and 29 rows of Midsummer, planted late in August. The infected plants were removed from a portion of each strip, 120 plants long, when the plants had recovered from transplanting. The increase in ClM was then recorded in the central part (50 plants long) of this area, and in an equal area which had not been rogued (Fig. 21). ClM incidence was similar in different parts of the long strips, so that errors from lack of randomization were small.

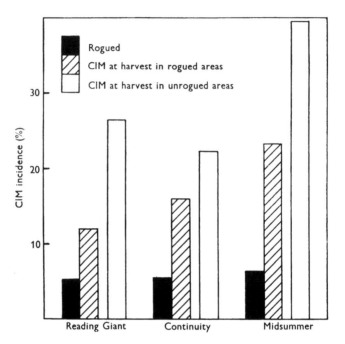

FIG. 21. Incidence of ClM in rogued and unrogued winter cauliflower crops, Datchet, 1952–3.

There was less ClM in the rogued than in the unrogued areas at harvest, the differences in incidence between them being 14·6 per cent. in Reading Giant, 6·3 per cent. in Continuity, and 16·4 per cent. in Midsummer (Table 28). About 2 per cent. more of the plants were lost by roguing than by disease (symptom categories (4)+(5)), but more plants were reduced in size

TABLE 28. Percentages of plants showing ClM at harvest in the different symptom categories in rogued and unrogued parts of crops, Datchet, 1952–3

Symptom category (see key in text)	Reading Giant		Continuity		Midsummer	
	Rogued	Unrogued	Rogued	Unrogued	Rogued	Unrogued
(1)	0·3	0·9	0·7	1·0	0·6	0·3
(2)	7·3	7·5	5·9	7·6	9·8	13·1
(3)	3·8	14·2	8·6	9·9	10·2	18·9
(4)	0·5	3·8	0·7	3·7	2·6	7·2
(5)	0	0·1	0	0	0	0·1
Rogued	5·4	.	5·5	.	6·4	.
Total	17·3	26·5	21·4	22·2	29·6	39·6

(category (3)) in the unrogued plots. The roguing was carefully done by trained persons, and could not be done so well by farmers. The results do not justify any recommendation that farmers would find roguing worth while.

Hertfordshire. The spread of ClMV and CBRSV in a summer cauliflower crop, variety Majestic, was recorded at Rothamsted. There were six replications of the following treatments, in plots 15 plants square:

(1) No 'infector' introduced.
(2) The central plant infected with ClMV.
(3) The central plant infected with CBRSV.
(4) The central plant infected with both viruses.

Winged *B. brassicae* were numerous from mid-July to the end of August (Table 25); some of them brought ClMV into the crop, and two kale crops on the farm had 5 and 18 per cent. of the plants infected at the end of September. ClMV spread widely throughout the experiment, and no differences were found between treatments; ClM incidence increased from 0·2 per cent. on 9 July to 24 per cent. in mid-August and 47 per cent. on 19 September. In contrast, CBRS increased from 0·2 to only 3·5 per cent. on 19 September; no CBRSV was introduced from outside the experiment, and the virus spread less readily within.

Curd diameter was measured in 50 plants each of four ClMV infection categories during the first two cuttings, on 29 September and 4 October; the mean diameters (in.) of 100 plants were: uninfected 6·0, late-infected 5·6, mid-season infected 5·0, early infected 4·4.

Hampshire, Kent and Warwickshire. Experiments were done by N.A.A.S. officers at Lymington, in the Isle of Thanet, and at Luddington, in which plants infected with ClMV were transplanted to the centres of plots, either soon after the crop was planted in July or early August, or later, and then spread of virus from these was recorded.

At Lymington aphids were few, and ClM incidence increased from 0·07 in July to 4·3 per cent. in November. About 230 plants were infected directly or

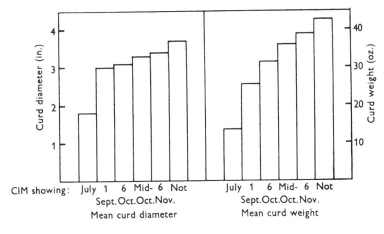

Fig. 22. Yield of cauliflower curds, variety Extra Early Roscoff, in relation to date of infection with ClMV, Lymington, 1952.

secondarily from the 5 'infectors' planted on 17 July, and about 60 from the 5 planted on 28 August. The variety was Extra Early Roscoff, and a large proportion of plants showing symptoms before October were stunted. The mean diameters and weights of uninfected and infected curds are shown in Fig. 22; curd diameter did not reflect the loss of yield caused by ClMV to the same extent as curd weight.

In Thanet a few *M. persicae* were trapped, but *B. brassicae* were fairly numerous during August and early September (Table 25). ClM incidence increased from 0·1 on 7 August to 4·0 per cent. in March. Virus spread from 'infectors' planted on 2 October was negligible; although most spread occurred in August and September, only 13 of the 137 newly infected plants showed ClM at the end of October, and 59 at the end of December.

At Luddington very large numbers of *B. brassicae* colonized the plants during early August, and brought ClMV with them. No newly infected plant was seen on 1 September, but by 15 September 40 per cent. of the crop showed ClM. Despite almost every plant being infected at harvest in April 1953, 80 per cent. of the curds were saleable—further evidence of the tolerance to infection of the variety Continuity.

Discussion of the results of the 1951–3 experiments

M. persicae and *B. brassicae* were the important vectors of ClMV; no evidence was obtained that any other species warranted consideration. A correlation between the numbers of *M. persicae* and *B. brassicae* trapped, and virus disease incidence, has not been established, however, possibly because the data obtained are very variable, covering different cauliflower varieties and different initial incidences of disease. Also the extent to which virus spreads varies with the activity of the aphids when they arrive in, or leave the crop, and this activity varies with different weather conditions; also the time they fly, in relation to the number of infected plants in the crop, affects virus spread, because an aphid's opportunity of feeding on an infected plant increases as the proportion of such plants increases during the season.

Although virus was often spread within crops after infected seedlings were transplanted, ClM incidence exceeded 40 per cent. at harvest only when incoming aphids brought virus with them (Luddington and Rothamsted, 1952). Most spread occurred during the period from transplanting to early October; very few plants were infected during the late autumn. Many of the plants infected during September, however, failed to show symptoms until spring. When plants were infected in the seedbed, they later were not only the source from which there was further spread, but were the plants which yielded least. Most of the spread within the crop was to plants near the 'infectors'.

It is postulated above that most virus spread within the crop is done by incoming winged aphids, but the habits of winged aphids when colonizing crops have not been adequately studied. Evidence from most studies of virus spread in field crops suggests that flights are short, e.g. Broadbent (1953) calculated that most flights in potato crops were over distances between 4 and 9 ft. Such short flights, when they are from infected to healthy plants, result in the 'pools' of infected plants that are so typical of virus diseases. These 'pools' have often been attributed mainly to virus spread by wingless aphids. As aphids can

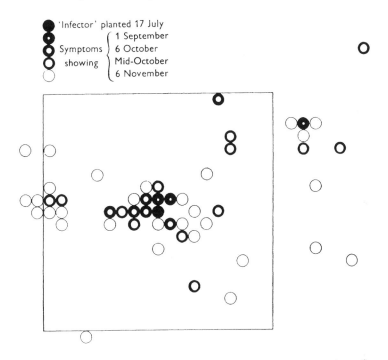

FIG. 23. Pattern of plants infected with ClMV around an 'infector' planted on 17 July 1952 at Lymington.

readily transmit ClMV after feeding for hours on an infected plant, unlike aphids that have fed long on plants infected with CBRSV, there is no reason why wingless aphids should not transmit ClMV during their wanderings; evidence was obtained in the seedbed experiments that CBRSV can also be spread by walking aphids. Nevertheless, much of the spread in the experiments was associated with aphid flights (see also Jenkinson, 1955); also many of the plants that were infected were not near infected plants (Figs. 20 and 23), and were almost certainly infected by winged aphids.

In seven experiments, where the sources of new infections could be identified with reasonable certainty, the mean lengths of the flights by the aphids responsible for virus transmission could be calculated, assuming that all such transmission was by winged aphids (Table 29). Most winter cauliflowers in the experiments were planted on the square, with 27 to 30 in. between plants. Distance can be estimated by drawing concentric circles around the 'infectors', but it was difficult to assign plants to definite annulae, and so plants in squares around the 'infectors' were used. The corner plants of a square are about 1·4 times as far away from the 'infector' as the plants in the middles of the sides, but there are the same numbers of plants in the two distances, so the flight lengths calculated are better referred to as 'plant-spaces' than as 'feet'. When the percentages of plants infected at any distance were high, they were transformed by the multiple-infection transformation (Gregory, 1948). The mean flight distances varied between 3·7 and 5·9 plant-spaces, with a mean flight length, from the 68 'infectors' involved, of 4·4 plant-spaces, i.e. about 11 ft. This will be a minimum figure, because in some experiments it was not possible to relate new infections to 'infectors' beyond seven to nine plants; there is also the possibility that some

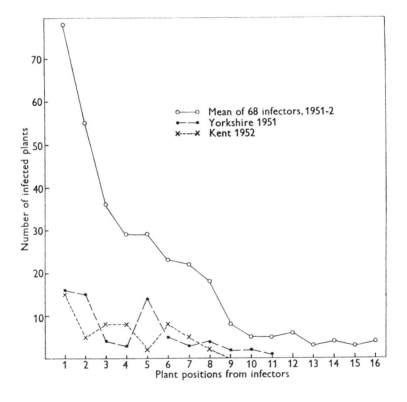

FIG. 24. Numbers of plants infected with ClMV at different distances (plant-spaces) from 'infector' plants.

TABLE 29. New CIMV infections at 1 to 16 plant-spaces from 'infectors', and calculated mean flight lengths of aphids

Site	No. of 'infectors'	Plant-spaces from 'infectors'																Mean flight length (plant-spaces)
		1	2	3	4	5	6	7	8	9	10	11	12	13	14	15	16	
Bucks. 1951	2	5	1	3	4	3	1	3	.	.	.	.	.	.	.	.	.	3·7
Cornwall 1952	1	2	3*	1	2	3	3	3	3	.	.	.	.	.	.	.	.	4·8
Hants. 1952	10	22*	18*	9	3	0	4	4	2	3	2	2	4	2	1	2	3	4·7
Kent 1951	15	5	5	1	2	1	0	2	1	2	1	2	2	0	1	1	1	5·9
Kent 1952	10	14*	5	8	8	2	8	5	2	0	0	0	0	1	2	.	.	4·4
Warwick. 1951	15	9	8	10	7	6	2	2	6	1	.	.	.	.	.	.	.	3·8
Yorks. 1951	15	15*	15	4	3	14	5	3	4	2	2	1	.	.	.	.	.	3·9
Total	68	72*	55	36	29	29	23	22	18	8	5	5	6	3	4	3	4	4·4

* Transformed before calculating flight length.

infections near an 'infector' might have been caused by aphids flying from distant 'infectors', or by walking aphids from a nearby one.

The experiments from which these data are compiled are few, but the graph (Fig. 24) of the summer infections (transformed), at different distances from the 68 'infectors', shows that 39 per cent. of the infections occurred within two plants of the 'infector', and most of the remainder (48 per cent. of total) were within the next six plants. In many of the experiments the number of infected plants falls with distance from the 'infectors' and then suddenly rises again. This is shown very clearly in the Yorkshire (1951) figures (Table 29) at the fifth plant, in the Kent (1952) figures at the sixth, and less clearly in the Hampshire (1952) figures at the sixth and seventh plants, and in the Buckinghamshire (1951) at the fourth. This may reflect a tendency on the part of the aphids to fly either a very short distance, thus infecting the first or second plant, or to make slightly longer flights over four to six plant-spaces. As flight distances would probably vary with the age of the aphid, and the weather, considerable variation might be expected from crop to crop, and in the same crop at different times. Further evidence that many flights are over distances of about four plant-spaces is provided by the results from Warwickshire (1951). Ten of the fifteen 'infectors' were the same age as the crop, but the other five were large old plants. The figures in Table 29 are the sum of infections around the fifteen 'infectors', but these are separated in Fig. 25, and show that there was

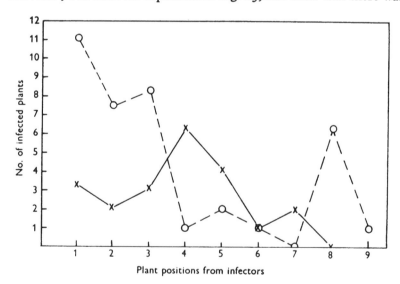

FIG. 25. Numbers of plants infected with ClMV at different distances from 'infectors' the same height as the crop (X ———— X), and taller than the crop (O — — — O), Warwickshire, 1951.

an increase in the number of infections at distances of four and five plant-spaces from the 'infectors' when these were the same age as the crop, and at a distance of eight plant-spaces when the 'infectors' were larger than the crop. Aphids usually take flight from the upper parts of plants, and it seems reasonable to suppose that they might fly farther, the higher the plant from which they took off in relation to surrounding plants. In the similar trial in Yorkshire the number of infections increased at the fifth plant-space, whether the 'infector' was old or the same age as the crop, but there was another increase at the tenth plant-space with old 'infectors', not shown with young ones.

The infections occurring in the plants adjacent to 'infectors' may be caused by winged aphids making short flights or 'false-starts', as they often do (Müller & Unger, 1951), or they might occasionally be caused by walking aphids.

Many more data of the type presented need to be collected before the hypothesis put forward can be confirmed or modified.

DIFFERENCES IN INCIDENCE OF CAULIFLOWER MOSAIC AND CABBAGE BLACK RING SPOT

During the years 1950–4 ClMV was much more prevalent in field crops than CBRSV; even when CBRSV was present it sometimes failed to spread, and when both viruses were deliberately introduced, it always spread less readily than ClMV.

CBRSV has a much wider host range than ClMV, which is confined to cruciferous plants; also sap from leaves infected with CBRSV has an infection end-point of about 1 in 10,000, in contrast to 1 in 2000 for sap from plants infected with ClMV. The results of transmission tests, given in Table 15, suggest that more aphid species transmit ClMV than CBRSV, but very few of the existing species have been tested, so it is premature to conclude that non-colonizing aphids may cause differences in spread, especially as the results of the field experiments suggest that *M. persicae* and *B. brassicae* are the only common vectors of both viruses. Nor can differences in the rate of spread of the two viruses be accounted for by differences between these vectors, which transmitted both viruses with equal facility in laboratory tests.

van Hoof (1954) and Hamlyn (1953, 1955) found that aphids could retain ClMV for longer than CBRSV, and also were infective after long periods of feeding on plants infested with ClMV, but not on plants infected with CBRSV. This is one reason why ClMV spreads more rapidly, but it may not be the main reason, because in an experiment during 1952, both viruses spread readily when the infected source plants were young but, when they were old, CBRSV spread less readily than ClMV (Broadbent, 1954). The rate at which the two viruses spread in the field could be accounted for by the different manner in which they are distributed in old infected plants, and the effect this has on transmission by aphids. ClMV occurs in high concentration in all the new leaves produced by infected plants; CBRSV, in contrast, occurs mainly in the older, lower leaves, and even there is localized in parts that show symptoms. Only in recently infected plants does CBRSV occur in young leaves. After flying, most aphids alight on the upper parts of plants; they are therefore more likely to acquire ClMV than CBRSV.

CONTROL MEASURES

The four years' studies have added many details to our knowledge about the epidemiology of cauliflower mosaic and cabbage black ring spot but little to the basic recommendations for control made by Prentice (1950). Control measures, as so often with virus diseases, remain largely matters of farm hygiene and common sense.

The most important control measure is to limit or prevent virus introduction into the seedbed, because losses are most severe when plants are infected at a young stage, and infected seedlings are the most important sources of virus spread within the crop. It has been shown that this can be done by isolation or, when this is not possible, by surrounding the seedbed with cereal barriers. Under no circumstances should a brassica seedbed be sown near other brassica crops. The seedbed should be sprayed with insecticide when aphids are very numerous—not to prevent the introduction of virus, but to prevent *B. brassicae* feeding damage and to limit virus spread within the bed. All seedlings should be examined when they are being transplanted; very large plants and those showing virus disease symptoms should be rejected, as should the outside rows of beds not surrounded by barriers. On present evidence the seedbed should not be more than twelve rows wide when barriers are used, and if wider should have barriers at twelve-row intervals. This recommendation might be modified in the light of later work. Perhaps the best barrier would be two or three rows of a mixture of barley and oats, because the barley grows quickly, but the oats grow taller. In long strip seedbeds an occasional drill of cereal across the drills would limit virus spread along the seedbed. Close spacing of the seedlings, so far as is consonant with the production of sturdy seedlings, will reduce disease incidence.

Whenever possible, brassica seedlings should not be transplanted to land near old maturing brassicas, because virus might be carried from the old crop into the new. An intervening crop, immune from the virus,

74

will often cause the aphids to lose their infectivity by feeding, or to fly elsewhere, and the larger such an intervening crop can be made, the less will be the risk of infection in the young crop. After harvest all crop residues should be destroyed as soon as possible, to prevent aphids on them flying away, or visiting aphids acquiring virus from them. Seed-beds should be destroyed also after the crop has been transplanted.

When aphids are numerous, and old crops on a farm are infected with virus, the 'build-up' of disease incidence in young crops might be limited by spraying the older crops with insecticide before the aphids leave them to carry virus elsewhere. Although no experiments have been done to test this, it seems likely that if most of the brassica crops could have been sprayed during September 1947, the serious outbreaks of cauliflower mosaic caused by *B. brassicae* in ensuing years might have been prevented.

Seed crops are undoubtedly one of the greatest sources of infection in brassica-growing areas, because they overlap in time with so many crops and seedbeds. Cauliflower mosaic incidence in Thanet, Kent, has been low since the growing of seed was stopped. But seed has to be grown somewhere, and the danger would be decreased if all seed-plants were treated periodically with systemic insecticides to prevent aphids multiplying on them and to kill aphids visiting them.

It has not been possible to test the use of insecticides on growing crops during this work, but the results of experiments on other crops suggest that it is doubtful if much control would be obtained with the brassica viruses, which can be spread quickly. Spread within the crop might be reduced slightly, but the main effect would be to prevent the development of winged aphids which take virus elsewhere.

Strict application of these control measures should prevent a high incidence of disease developing in an area, but much will depend on the co-operation of neighbouring farmers, especially in areas of intensive cultivation such as Thanet, Evesham, Melbourne and Wakefield, where brassicas are grown as market-garden crops and overlap in time. When land is never free from brassicas the virus can be spread at any time from March to November. The diseases are scarce in Cornwall, despite the large acreage of winter cauliflower, because few other brassica crops are grown, even in allotments, and harvest is early before seedbeds are sown. Control will be much easier in areas where brassicas are treated as one of the crops in the general agricultural rotation, as in Devon. If brassica crops on such farms were to become seriously in-fected, the quickest method of reducing the incidence of disease would be to omit one crop, so that the farm was free from brassicas for a short period.

Allotments and gardens are frequent sources of virus and aphids, and growers near towns will no doubt always find it difficult to prevent the introduction of virus from these sources.

The effectiveness of roguing the transplanted crop will depend upon the time of virus spread, and the amount of later spread into the crop. When spread within the crop occurs soon after transplanting, the removal of infected plants in late July or early August will have little effect because each will probably have been the source of a number of new infections, and these plants will soon be infective themselves. In a year when most spread is in late August or September, however, roguing might be worth while.

Ultimately the most successful measure of control will be to plant resistant or tolerant varieties. Existing varieties differ widely in susceptibility and tolerance to infection, and breeding and selecting should be able to produce resistant varieties.

5. TURNIP YELLOW MOSAIC

The first virus transmitted by biting insects to be described in Britain was turnip yellow mosaic (Markham & Smith, 1949). Grasshoppers, earwigs and mustard beetles transmitted the virus in glasshouse experiments, but flea-beetles are the common vectors in the field. Markham & Smith stated that the virus was apparently confined to the Cruciferae, and it had been found occurring naturally in turnips, swedes and winter cauliflower. They infected cabbage, radish, kohlrabi, watercress, and several cruciferous weeds.

The Department of Agriculture for Scotland (*in litt.*) reported that the disease has been common in experimental turnip plots at Corstorphine, Edinburgh, and had been reported in winter cauliflower in East Lothian (1947) and in turnips in Fife (1951). Seed from infected turnips from Fife gave virus-free seedlings.

Turnip plants infected with turnip yellow mosaic virus (TYMV) were found during 1954 at Udny in Aberdeenshire and Mylnefield, near Dundee. A variant of TYMV was isolated from Honesty (*Lunaria annua* L.) plants growing in a garden near Bristol, near where the disease was once found in a turnip crop.

Croxall, Gwynne & Broadbent (1953) described what appeared to be TYM in crops of winter cauliflower, savoy, cabbage and brussels sprouts near the coast of Northumberland and Durham. This disease had been noted by some farmers in previous years, but was thought to be ClM. However, during 1952 there was a serious outbreak of the disease on most brassica-growing farms in a small area, coincident with exceptionally large numbers of flea-beetles, and during the ensuing years the virus was spread over a much wider area of the north-east.

The Northumberland virus is serologically related to TYMV, but differs enough antigenically for the two strains to be distinguished (Blencowe & Nixon, 1957). The two strains also cause different symptoms in many host plants (Broadbent & Heathcote, 1957): the Northumberland strain readily infects cauliflower (Pl. 7c), brussels sprouts and cabbage, and causes obvious symptoms, whereas these plants are not very susceptible to the Edinburgh strain, and when infected are almost symptomless. Turnip (Pl. 7a), Chinese cabbage, swede, charlock (Pl. 7b), shepherd's purse and many other cruciferous weeds are susceptible to both strains, but react differently to them.

Infection with one strain does not protect plants against infection with the other. Both strains are inactivated by heating sap for 10 min. at temperatures between 75 and 80° C. The infection end-point for both strains is also about 10^5.

Both strains were transmitted by flea-beetles but not by the aphids *Myzus persicae* and *Brevicoryne brassicae*. The virus was occasionally transmitted from infected to healthy turnip plants in a wind tunnel, so field spread by contact in windy weather may be possible.

The Edinburgh strain, though apparently widely distributed in Scotland and in some parts of England, probably overwintering in weeds, has so far caused little loss, as only a few turnip crops have been severely affected. The Northumberland strain, however, causes more severe losses than cauliflower mosaic in affected crops, but fortunately it is still confined to two counties.

As flea-beetles are relatively easy to control with insecticides, the spread of these viruses may be more easily checked than that of the aphid-transmitted viruses. Experiments to test this are being made in Northumberland and Durham. It is too early to make definite recommendations, but rigorous flea-beetle control in seedbeds, coupled with seedbed isolation, and with dusting or spraying of infected crops and surrounding hedgerows, ought to be effective.

6. TURNIP CRINKLE

Another flea-beetle-transmitted virus that severely affects turnip is turnip crinkle (Pl. 8), discovered on a farm in Kincardineshire in 1953, and so far not found elsewhere (Broadbent & Heathcote, 1957). It is not

77

serologically related to TYMV (Blencowe & Nixon, 1957), but like that virus, occurs in large amounts in the sap of infected plants (withstanding dilution to 10^{-5}), and is inactivated after being heated for 10 min. at temperatures between 80 and 85° C.

The virus was readily transmitted from infected to healthy turnip plants in a wind tunnel, so field spread by contact in windy weather is likely.

Turnips, swedes and charlock were infected in the field. Turnips show two types of symptoms; one, crinkling of the leaves, with the margins curled inwards, and indistinct, irregular, light green or yellow patches on the slightly stunted plants; two, the leaves were distorted and crinkled, and the plants severely stunted and rosetted. Turnip seedlings infected with sap from rosetted plants usually died. Swedes, cauliflowers, brussels sprouts and cabbages were infected, but showed mild or no symptoms. Many cruciferous weeds were readily susceptible, but whereas some, such as charlock (Pl. 7b), were severely distorted, others such as shepherd's purse showed only a slight mottle and leaf crinkle. The virus is not confined to Cruciferae; although *Nicotiana* spp. were not susceptible, the virus caused faint chlorotic lesions in *Datura stramonium* L. and became systemic, but non-inoculated leaves were symptomless; necrotic local lesions developed on leaves of *Gomphrena globosa* L. and *Chenopodium amaranticolor* Coste & Reyn. rubbed with sap from infected plants.

7. CUCUMBER MOSAIC

The only other virus known that occasionally infects brassica crops in Britain is cucumber mosaic virus, which is transmitted by aphids. One Rothamsted strain that was inoculated to cabbage, cauliflower and brussels sprouts multiplied in the plants without causing symptoms; turnips and swedes were slightly mottled and had wrinkled leaves, somewhat similar to plants infected with a mild strain of turnip crinkle virus.

8. SUMMARY

1. Viruses common in British brassica crops are cauliflower mosaic and cabbage black ring spot, both transmitted by aphids. Their presence in Britain was first recorded in 1934, but it is likely that they were present many years before. Other viruses, locally severe in the north, are turnip yellow mosaic and turnip crinkle; these are transmitted by flea-beetles and other biting insects.

2. The more important brassica-growing areas are in the eastern half of England and in the south-west of England and Wales. The total brassica acreage increased by about one-half during the 1939–45 war, mainly because of an increase in agricultural brassica crops for stock feeding. The horticultural brassica acreage increased rapidly after the war, in 1948 being double that in 1941. The principal centres of cauliflower growing are in counties with high proportions of brassica crops relative to other crops. The acreage of cauliflower was reduced considerably after 1948, partly because of severe outbreaks of cauliflower mosaic.

3. The properties of cauliflower mosaic virus are: (a) its thermal inactivation temperature is slightly above 70° C.; (b) its infection end-point is about 1 in 2000; (c) it withstands storage for about 6 days at 20° C.; (d) its host range is limited to Cruciferae; (e) symptoms are more severe at temperatures below 24° C.

The properties of cabbage black ring spot virus are: (a) its thermal inactivation temperature is slightly below 65° C.; (b) its infection end-point is about 1 in 10,000; (c) it withstands storage for about 2 days at 20° C.; (d) its host range is not limited to the Cruciferae; (e) symptoms are more severe at temperatures above 24° C.

4. Cauliflower mosaic symptoms include vein-clearing, vein-banding, leaf-distortion, necrotic stipple, and stunting. Cauliflower, Hungry Gap kale, turnip, swede and mustard react severely when infected; cabbage, brussels sprout and kales are more tolerant.

Cabbage black ring spot symptoms in cauliflower consist of chlorotic or necrotic spots or rings, usually discrete and raised slightly to form blisters. Turnip, swede and mustard react severely when infected, cabbage and brussels sprouts react moderately severely, and cauliflower and kale are usually more tolerant.

Plants infected simultaneously with both viruses usually show a generalized mottle, and react more severely than plants infected with one virus.

With both viruses, different varieties of cauliflower and other hosts vary in susceptibility to infection and tolerance when infected. Young plants are more easy to infect, and symptoms appear sooner than in old plants.

5. Both cauliflower mosaic and cabbage black ring spot viruses occur in a number of strains. Mild strains of both viruses may not produce recognizable symptoms in field cauliflowers, and only moderately distort turnips; severe strains kill turnips.

6. Symptom development depends on temperature and on the rate of growth of the plants. Young plants infected in summer show cauliflower

mosaic symptoms in four to six weeks; plants infected in autumn fail to show symptoms until spring. Under glass, symptoms show in 2–4 weeks according to the temperature; the corresponding time for cabbage black ring spot is $1\frac{1}{2}$–3 weeks. Cauliflower mosaic symptoms are masked when plants grow at temperatures over 24° C., and often also during winter if plants are infected with mild strains. Shading delays the appearance of symptoms in both diseases, and diminishes their severity.

7. In manurial experiments, nitrogenous fertilizer significantly increased the rate at which symptoms developed and also their severity; potash had no significant effect, but phosphate significantly depressed symptoms, especially in the absence of potash.

High nitrogen increased plant susceptibility to ClMV. Virus infection decreased the size, weight and quality of curds, especially of plants infected early. Infection increased the proportion of bracty curds.

High levels of nitrogen, especially given as dung or hoof and horn meal, increased the incidence of 'tip-burn'; more plants of the varieties St George and Continuity died during the winter because of high nitrogen than because of ClMV infection. Large quantities of the organic manures also decreased yield, and moderate amounts of dung and nitro-chalk together gave maximum yields.

Seedlings transplanted late to fill gaps usually failed to produce marketable curds.

8. Both viruses move from inoculated leaves and reach the growing point many days before symptoms appear. Fully expanded leaves are not infected, but partially expanded leaves and newly developing leaves are. Cauliflower mosaic virus is present in the curd and flowers of infected cauliflower plants.

As the plant grows, cabbage black ring spot virus, unlike cauliflower mosaic virus, becomes more concentrated in the older leaves, being often undetectable in the young ones. This concentration gradient is reflected in the situation and severity of cabbage black ring spot symptoms, which are confined mainly to the older leaves. Cabbage black ring spot virus has not been detected in the curds and flowers of infected cauliflower plants. The concentration of cabbage black ring spot virus increases for some weeks after infection and then decreases as the plant ages.

9. Although many common aphids can transmit one or both viruses field experiments and observations suggest that *Myzus persicae* (Sulzer) and *Brevicoryne brassicae* (L.) are the only important vectors. Most serious outbreaks of cauliflower mosaic have recently been associated with the presence of many *B. brassicae*, an aphid which breeds only on

crucifers, and is most plentiful in summer soon after winter cauliflowers are transplanted. *M. persicae* is usually more numerous than *B. brassicae* in spring and is important for spreading virus from old crops to seedbeds and newly planted crops. Sometimes it is the main vector during the summer and autumn.

10. Periods of fasting before feeding on the infected plant did not increase the capacity of vectors of cauliflower mosaic virus to acquire virus, and it mattered little if they fed for a short or long time on the source. Virus survived for at least three hours in feeding aphids, and in fasting aphids it occasionally survived for 24 hours.

Vectors acquired cabbage black ring spot virus more readily after at least 15 minutes' absence from a food plant. Ten seconds feeding on an infected plant was sufficient to enable them to acquire virus, and feeding for more than 15 minutes considerably reduced the infectivity of the aphids, unless they were constantly moving and probing the leaf. Having acquired virus, aphids transmitted it best if moved to a healthy plant immediately. Virus sometimes survived in feeding aphids for half an hour, but in fasting aphids it occasionally survived for nine hours.

11. Winged aphids, bred on infected cauliflower plants under cages, were transferred to seedlings after they had flown from the plants to find out what proportion of them were infective. Fifteen to 20 per cent. of those bred on plants of different ages infected with cauliflower mosaic virus transmitted virus, about the same proportion as those bred on young plants infected with cabbage black ring spot virus; but aphids bred on older plants infected with cabbage black ring spot virus infected few seedlings.

12. Cauliflower mosaic is usually more prevalent in field crops than cabbage black ring spot, despite the wider host range and other pro-perties of cabbage black ring spot virus. This is because (*a*) aphids can retain cauliflower mosaic virus longer than cabbage black ring spot virus, and (*b*) whereas cauliflower mosaic virus occurs in high concentra-tion in all new leaves developed by infected plants, cabbage black ring spot virus occurs mainly in the older, lower leaves. Only in recently infected plants can cabbage black ring spot virus be detected in young leaves. After flying, most aphids alight on the upper parts of plants; they are therefore more likely to acquire cauliflower mosaic virus than cabbage black ring spot virus.

13. Virus can be spread rapidly from a few infected plants within the crop, so that a crop planted with about 0·5 per cent. of the seedlings infected may have 20–40 per cent. of the plants infected at harvest; the more seedlings that are infected when transplanted, the greater will be the final incidence of disease. Most spread occurs before early October,

but many plants that are infected during September or later fail to show symptoms until spring, and their yield is then little reduced. Cauliflower plants infected during July and August usually show symptoms during the autumn, and either produce smaller curds or no saleable curds, depending on the variety. Most of the spread is to plants within eight plant-spaces of the virus source, and these plants then form sources for further spread. Winged aphids are mainly responsible for spreading virus. Removing infected plants soon after transplanting resulted in fewer infected plants in the rogued crop, but the labour was not worth while in a year when much spread occurred before the plants were rogued.

14. Disease incidence is usually greatest when the seedbed has been near an old infected crop. Uninfected seedlings can be grown if the seedbed is placed away from overwintering brassica crops. Relative isolation on a large farm is often just as effective as isolation by a mile or more from known sources.

15. More plants are obtained per pound of seed by drilling than by broadcasting; the percentage of plants infected with virus (but not the total number) will be lower if drills are put closer together; as the seed-rate is increased the percentage of infected plants will be lowered, but the percentage of seedlings suitable for transplanting will be lowered also; 15 to 20 plants per foot of row appears to be about optimal; the largest plants are the most likely to be infected and should be rejected when transplanting; more plants are infected in outer rows of seedbeds than in inner rows, and if possible outside rows should not be transplanted.

16. Spraying seedbeds with various contact and systemic insecticides did not significantly decrease disease incidence. Insecticides might be useful, however, if aphids were very numerous, to prevent both *B. brassicae* feeding damage and secondary virus spread within the bed.

17. Plant barriers around seedbeds reduced virus disease incidence in the seedlings to about one-fifth that in unprotected beds. Cereals were the best barrier plants, even single rows being effective. This method of protection would be particularly useful on small holdings where the seedbed cannot be sown far from standing crops.

18. Virus is often spread from old crops to nearby young ones, especially to plants in the outer rows nearest to the old crop. Such gradients of infection are not found when spread is mainly from sources within the crop, unless aphids are more numerous at one side of the crop than at the other. Crop to crop spread on farms can be reduced by planting crops of similar age in large blocks (instead of in small scattered areas), separated from older plantings by as great a distance as possible.

19. Turnip yellow mosaic, a virus transmitted by flea-beetles and some other biting insects, exists in several strains. One strain affects turnips, mainly in Scotland, and has little effect on other cultivated brassicas. Another strain, at present confined to north-east England, causes severe losses in cauliflower, cabbage and other crops. Many cruciferous weeds are susceptible to both strains. The virus is highly concentrated in plant sap, and may be transmitted by plant contact in windy weather.

20. Turnip crinkle, a virus disease so far found only on one farm in Scotland, is transmitted by flea-beetles, and by contact. Infected turnips and some cruciferous weeds are severely stunted or killed, but other cultivated brassicas are almost symptomless.

21. Cucumber mosaic virus causes slight distortion of turnip leaves, but is carried without symptoms in most other cultivated brassicas.

ACKNOWLEDGEMENTS

Numerous people have co-operated in the work described in this report, and I wish to thank particularly the following for their generous help and advice: The Directors and Technical Officers of the N.A.A.S. Experimental Horticulture Stations, viz: Messrs P. H. Brown & G. F. Wheeler (Efford); Messrs H. Fairbank & N. Moss (Luddington); Messrs F. G. Smith & A. Brown (Stockbridge House). The Plant Pathologists of the N.A.A.S., in particular: Messrs W. Buddin & E. T. Roberts (Reading); Drs H. H. Glasscock & A. G. Walker, Miss U. E. K. Fraenkel & Mr J. J. Walker (Wye); Messrs L. Ogilvie & A. G. Robertson (Bristol); Messrs G. H. Brenchley & W. Campbell (Starcross); Mr E. Lester (Derby); Messrs N. C. Preston & W. R. Rosser (Wolverhampton); Dr E. Taylor, Mr J. E. E. Jenkins & Miss E. R. Schofield (Evesham); Dr I. F. Storey, Mr J. P. Cleary & Miss A. E. Godwin (Leeds); Dr H. E. Croxall & Mr D. C. Gwynne (Newcastle). The Entomologists of the N.A.A.S., in particular: Messrs B. D. Moreton & W. I. StG. Light (Wye); Mr B. A. Cooper (Kirton); Mr C. A. Collingwood (Evesham) & Mr H. G. Morgan (Starcross). Messrs C. G. Finch, V. F. Grainger, K. E. Haine, P. D. Broad, C. North, G. H. Daniel and W. E. H. Fiddian of the National Institute of Agricultural Botany; Dr I. W. Prentice (Ministry of Agriculture); Dr K. M. Smith (Molteno Institute, Cambridge); Mr J. G. Jenkinson (Seale-Hayne Agricultural College, Devon); Drs T. P. McIntosh, C. E. Foister, Elizabeth G. Gray and C. H. Cadman (Scotland); Messrs Barker Bros., Riding Court Farm, Datchet, Bucks., and other farmers who kindly allowed observations to be made on their crops; the staffs of the Rothamsted Farm, Plant Pathology, Insecticides, Statistics and Field Experiments Departments, in particular Dr T. W. Tinsley, Mr G. D. Heathcote and Mr J. H. A. Dunwoody; Mr V. Stansfield, Rothamsted, for photography and the numerous N.A.A.S. Officers and others who have changed the aphid traps.

REFERENCES

BLENCOWE, J. W. & TINSLEY, T. W. (1951). The influence of density of plant population on the incidence of yellows in sugar beet crops. *Ann. appl. Biol.* **38**, 395–401.

BLENCOWE, J. W. & NIXON, H. L. (1957). Studies on crucifer viruses transmitted by flea beetles. 2. (In preparation.)

BROADBENT, L. (1950). The correlation of aphid numbers with the spread of leaf roll and rugose mosaic in potato crops. *Ann. appl. Biol.* **37**, 58–65.

BROADBENT, L. (1953). Aphids and virus diseases in potato crops. *Biol. Rev.* **28**, 350–80.

BROADBENT, L. (1954). The different distribution of two brassica viruses in the plant and its influence on spread in the field. *Ann. appl. Biol.* **41**, 174–82.

BROADBENT, L., DONCASTER, J. P., HULL, R. & WATSON, M. A. (1948). Equipment used for trapping and identifying alate aphides. *Proc. R. ent. Soc. Lond.* (A) **23**, 57–8.

BROADBENT, L. & HEATHCOTE, G. D. (1955). Sources of overwintering *Myzus persicae* (Sulzer) in England. *Plant Path.* **4**, 135–7.

BROADBENT, L. & HEATHCOTE, G. D. (1957). Studies on crucifer viruses transmitted by flea-beetles. 1. (In preparation.)

BROADBENT, L. & TINSLEY, T. W. (1953). Symptoms of cauliflower mosaic and cabbage black ring spot in cauliflower. *Plant Path.* **2**, 88–92.

BROADBENT, L., TINSLEY, T. W., BUDDIN, W. & ROBERTS, E. T. (1951). The spread of lettuce mosaic in the field. *Ann. appl. Biol.* **38**, 689–706.

CALDWELL, J. & PRENTICE, I. W. (1942*a*). A mosaic disease of broccoli. *Ann. appl. Biol.* **29**, 366–73.

CALDWELL, J. & PRENTICE, I. W. (1942*b*). The spread and effect of broccoli mosaic in the field. *Ann. appl. Biol.* **29**, 374–9.

CROXALL, H. E., GWYNNE, D. C. & BROADBENT, L. (1953). Turnip yellow mosaic in broccoli. *Plant Path.* **2**, 122–3.

DENNIS, R. W. G. & FOISTER, C. E. (1941). List of diseases of economic plants recorded in Scotland. *Trans. Brit. mycol. Soc.* **25**, 266–306.

EMPSON, D. W. (1952). Survey of cabbage aphid populations on brussels sprouts. 1946–51. *Plant Path.* **1**, 36–8.

GLASSCOCK, H. H. & MORETON, B. D. (1955). Cauliflower mosaic in East Kent. *Agriculture*, **62**, 270–4.

GREGORY, P. H. (1948). The multiple-infection transformation. *Ann. appl. Biol.* **35**, 412–17.

HAMLYN, B. M. G. (1952). Studies on the virus diseases of brassicas. M.Sc. Thesis, University of Reading.

HAMLYN, B. M. G. (1953). Quantitative studies on the transmission of cabbage black ring spot virus by *Myzus persicae* (Sulz.). *Ann. appl. Biol.* **40**, 393–402.

HAMLYN, B. M. G. (1955). Aphid transmission of cauliflower mosaic. *Plant Path.* **4**, 13–16.

HEATHCOTE, G. D. (1955). The behaviour of aphids and its effect upon the spread of plant virus diseases. M.Sc. Thesis, University of London.

HOOF, H. A. VAN (1952). Stip in Kool, een virus ziekte. *Meded. Dir. Tuinb.* **15**, 727–42.

HOOF, H. A. VAN (1954). Verschillen in de overdracht van het bloemkool-mozaiekvirus bij *Myzus persicae* Sulzer en *Brevicoryne brassicae* L. *Tijdschr. Pl. Ziekt.* **60**, 267–72.

JENKINSON, J. G. (1955). The incidence and control of cauliflower mosaic in broccoli in south-west England. *Ann. appl. Biol.* **43**, 409–22.

JENKINSON, J. G. & GLYNNE JONES, G. D. (1951). Control of cauliflower mosaic virus. *Nature, Lond.*, **168**, 336–7.

JENSEN, M. (1954). *Shelter Effect*. Copenhagen: Danish Technical Press.

KVICALA, B. A. (1945). Selective power in virus transmission exhibited by an aphis. *Nature, Lond.*, **155**, 174–5.

KVICALA, B. A. (1948a). Studies on the composite nature of cauliflower mosaic with special reference to the selective transmission of both viruses by certain aphids. *Acta Univ. Agric. Silv., Brunn*, **40**, 1–87.

KVICALA, B. A. (1948b). A virus mosaic of cabbage and the relationship to the aphids. *Ann. Acad. tchecosl. Agric.* **21**, 32–42.

MARKHAM, R. & SMITH, K. M. (1946). A new crystalline plant virus. *Nature, Lond.*, **157**, 300.

MARKHAM, R. & SMITH, K. M. (1949). Studies on the virus of turnip yellow mosaic. *Parasitology*, **39**, 330–42.

MARKKULA, M. (1953). Biologisch-ökologische Untersuchungen über die Kohlblattlaus, *Brevicoryne brassicae* (L.). *Ann. Zool. Soc. 'Vanamo'*, **15**, 1–113.

McCLEAN, A. P. D. & COWIN, S. M. (1952–3). Diseases of crucifers and other plants caused by cabbage black ring spot virus. *Sci. Bull. Dep. Agric. S. Afr.* 332.

MILLARD, W. A. (1945). Canker and mosaic of broccoli. *J. Minist. Agric.* **52**, 39–42.

MOERICKE, V. (1951). Eine Farbfalle zur Kontrolle des Fluges von Blattläusen, insbesondere der Pfirsichblattlaus, *Myzodes persicae* (Sulz.). *NachrBl. dtsch. PflSchDienst.*, Stuttgart, **3**, 23–4.

MOORE, W. C. (1943). Diseases of crop plants. *Bull. 126 Min. Agric. & Fish.* London: H.M.S.O.

MOORE, W. C. (1948). Diseases of crop plants. *Bull. 139 Min. Agric. & Fish.* London: H.M.S.O.

MÜLLER, H. J. & UNGER, K. (1951). Über die Ursachen der unterschiedlichen Resistenz von *Vicia faba* L., gegenüber der Bohnenblattlaus *Doralis fabae* Scop. *Züchter*, **21**, 1–30.

OGILVIE, L. (1934). *Rep. agric. hort. Res. Sta. Bristol*, 1934, 180.

PLANK, J. E. VAN DER (1947). The relation between the size of plant and the spread of systemic diseases. *Ann. appl. Biol.* **34**, 376–87.

POUND, G. S. (1946). Control of virus diseases of cabbage seed plants in western Washington by plant bed isolation. *Phytopathology*, **36**, 1035–9.

POUND, G. S. & WALKER, J. C. (1945 a). Differentiation of certain crucifer viruses by the use of temperature and host immunity reactions. *J. agric. Res.* 71, 255–78.

POUND, G. S. & WALKER, J. C. (1945 b). Effect of air temperature on the concentration of certain viruses in cabbage. *J. agric. Res.* 71, 471–85.

PRENTICE, I. W. (1950). Broccoli mosaic. *Agriculture*, 56, 577–9.

SCHULTZ, E. S. (1921). A transmissible mosaic disease of Chinese cabbage, mustard and turnip. *J. agric. Res.* 22, 173–7.

SEVERIN, H. H. P. & TOMPKINS, C. M. (1948). Aphid transmission of cauliflower mosaic virus. *Hilgardia*, 18, 389–404.

SMITH, K. M. (1935). A virus disease of cultivated crucifers. *Ann. appl. Biol.* 22, 239–42.

SMITH, K. M. (1937). *A Textbook of Plant Virus Diseases*. London: Churchill.

SMITH, K. M. (1945). *Virus Diseases of Farm and Garden Crops*. Worcester Press.

STOREY, I. F. & GODWIN, A. E. (1953). Cauliflower mosaic in Yorkshire, 1950–51. *Plant Path.* 2, 98–101.

SYLVESTER, E. S. (1953). Host range and properties of *Brassica nigra* virus. *Phytopathology*, 43, 541–6.

TOMPKINS, C. M. (1937). A transmissible mosaic disease of cauliflower. *J. agric. Res.* 55, 33–46.

WALKER, J. C., LeBEAU, F. J. & POUND, G. S. (1945). Viruses associated with cabbage mosaic. *J. agric. Res.* 70, 379–404.

WATSON, M. A. & ROBERTS, F. M. (1939). A comparative study of the transmission of Hyoscyamus virus 3, potato virus Y and cucumber virus 1 by the vectors *Myzus persicae* (Sulzer), *M. circumflexus* (Buckton) and *Macrosiphum gei* (Koch). *Proc. Roy. Soc.* B, 127, 543–76.

(a)   *(b)*

(c)   *(d)*

PLATE I. SYMPTOMS OF CAULIFLOWER MOSAIC ON CAULIFLOWER
(*a*) Vein-clearing; (*b*) vein-banding; (*c*) necrotic areas or 'enations' (highly
magnified); (*d*) necrotic 'stipple'.

PLATE 2. SYMPTOMS OF CAULIFLOWER MOSAIC

(a) Young cauliflower plant recently infected; (b) cauliflower plants, variety Majestic, infected (*right*) and uninfected (*left*); (c) turnip killed by ClMV; (d) Dutch savoy cabbage showing 'stipple'.

(a)

(b)   (c)

PLATE 3. CABBAGE BLACK RING SPOT SYMPTOMS

(a) On Extra Early Roscoff cauliflower leaves (transmitted light); (b) and (c) on
purple sprouting broccoli leaves; (b) upper surface; (c) lower surface.

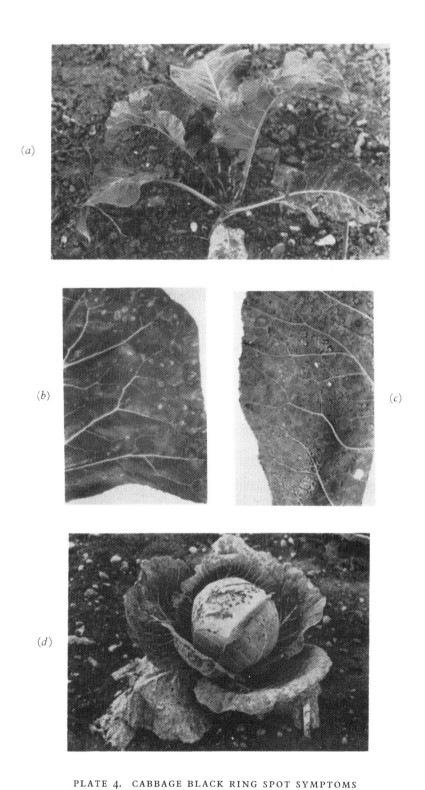

PLATE 4. CABBAGE BLACK RING SPOT SYMPTOMS

(a) On cauliflower, variety Majestic;   (b) on Roscoff X 5 leaf (upper surface);
(c) on Extra Early Roscoff leaf (lower surface)—severe strain;   (d) on cabbage,
variety Primo.

(a) (b)

(c) (d)

PLATE 5

(a), (b) Symptoms caused by a mixture of cauliflower mosaic and cabbage black ring spot viruses in cauliflower; (c) spots due to fungal attack on aphids on a cauliflower leaf; (d) damage caused by *Brevicoryne brassicae* feeding.

(a)

(b)

PLATE 6

(a) Cages used to determine the proportion of infective aphids bred on infected plants; (b) a flat sticky trap for aphids, and box for transporting the celluloid covers.

(a)

(b)

(c)

PLATE 7

(a) Virus disease symptoms on turnip. *Left to right*: uninfected, turnip yellow mosaic (Northumberland strain), TYM (Bristol strain), TYM (Edinburgh strain), turnip crinkle; (b) virus disease symptoms on charlock (*Sinapis arvensis* L.). *Left to right, upper row*: turnip crinkle, young and old infections, cauliflower mosaic, cabbage black ring spot; *lower row*: turnip yellow mosaic (Edinburgh strain), TYM (Northumberland strain), TYM (Bristol strain), uninfected; (c) turnip yellow mosaic (Northumberland strain) on Extra Early Roscoff cauliflower.

(a)

(b)

PLATE 8. TURNIP CRINKLE ON TURNIP
(a) Moderately severe isolate; (b) severe isolate.

INDEX

Aberdeen, 7, 76
Abutilon theophrasti, 35
Acreages of brassica crops, 2, 40, 79; Figs. 1, 2, 3
Allotments, 2, 60, 76
Alpine rock-cress, 35
Aphids, 21, 27, 28, 31
    alate movement, 16, 38, 39, 50, 58, 60, 61, 63, 65, 68, 74; Figs. 15, 24, 25
    apterous movement, 39, 52, 69, 73
    feeding damage, 15, 46, 52, 53, 74, 82; Pl. 5d
    overwintering, 39, 41
    predators and parasites, 15, 42
    virus transmission, 1, 6, 14, 17, 32, 36, 38, 42, 49, 58, 61, 68, 73, 78, 80; Pl. 6a; T. 15, 16, 18
Aphid-proof cages, 17, 44, 58
Aphid traps
    sticky cylindrical, 21, 37, 39; T. 16, 19, 25
    sticky flat, 27, 28, 31, 39, 56, 65; Pl. 6b; T. 17, 22
    water, 39
Arabis alpina, 35
Armoracia rusticana, 34
Ayrshire, 7

Barbados gooseberry, 35
Barbarea verna, 34
B. vulgaris, 35
Barker Bros., Datchet, Bucks., 84
Barley seedbed barriers, 52, 74; Fig. 18; T. 20, 21, 23, 24
Barrier crops, 49, 50, 74, 82; Figs. 15, 16, 17, 18; T. 20, 21, 23, 24
Berteroa incana, 35
Beta vulgaris, 35
Blencowe, J. W., 49, 77, 78
Bracts in cauliflower curds, 8, 22, 27, 29, 31, 80
Brassica campestris, 34
B. chinensis, 34
B. juncea, 34
B. napo-brassica, 34; T. 1
B. napus, 34; T. 1
B. nigra, 13, 34
B. oleracea
    var. *acephala*, 34; T. 1

B. oleracea—cont.
    var. *botrytis*, 34; T. 1
    var. *bullata*, 34
    var. *capitata*, 34; T. 1
    var. *caulorapa*, 34; T. 1
    var. *gemmifera*, 34; T. 1
    var. *italica*, 34
B. rapa, 34; T. 1
Brenchley, G. H., 84
Brevicoryne brassicae, 36, 73, 77, 80
    feeding damage, 15, 46, 52, 53, 74, 82; Pl. 5d
    incidence, 21, 27, 28, 31, 38, 41, 55, 56, 57, 60, 63, 65, 67, 68; T. 16, 17, 25
    infectivity of alatae from infected plants, 42, 81; Pl. 6a; T. 18
    overwintering, 41
    retention of virus, 38, 55
Bristol, 7, 76
Broad, P. D., 84
Broad bean seedbed barrier, 55; T. 23, 24
Broadbent, L., 7, 8, 21, 33, 39, 41, 58, 69, 74, 76, 77; T. 15
Broccoli, *see* Sprouting broccoli *and* Winter cauliflower
Brown, A., 84
Brown, P. H., 55, 58, 84
Brussels sprouts, 1, 2, 6, 9, 34, 39, 40, 43, 76, 77, 78, 79; T. 1
Buckinghamshire, 58, 60, 63, 65, 72; Figs. 19, 20, 21; T. 17, 25, 26, 27, 28
Buddin, W., 55, 58, 84

Cabbage, 1, 2, 6, 9, 16, 34, 39, 43, 59, 76, 77, 78; Fig. 19; Pl. 2d, 4d; T. 1
Cabbage black ring spot
    chlorotic spots and rings, 9, 79; Pl. 3a, b, c
    control by plant barriers round seedbeds, 50; Fig. 16; T. 20
    effect of plant growth on symptoms, 9, 33; Pl. 4a; T. 14
    effect of weather on symptoms, 16, 20, 80; Fig. 6; T. 4
    flower 'break', 13
    incidence, 6, 41, 73, 81; T. 16
    leaf distortion, 9, 13
    local lesions on tobacco, 21, 32, 33; T. 14

Cauliflower mosaic virus—*cont.*
  effect on seed yield, 60
  effect on yield of cauliflower curds, 23,
    26, 27, 29, 31, 67, 68, 80, 82; Figs. 8,
    9, 22; T. 9
  host plants, 1, 8, 9, 34, 79
  in cauliflower curds and flowers, 32, 60,
    80
  infection end-point, 7, 73, 79
  inoculation, 14, 32; T. 2, 3, 4
  killing plants, 6, 8, 9, 14, 22, 29, 79;
    Pl. 2c; T. 1
  sources of virus, 58, 59, 60, 61, 74, 82
  spread from crop to crop, 40, 49, 50,
    58, 61, 67, 68, 74, 82; Fig. 19
  spread to seedbeds, 12, 40, 44, 59, 74,
    82
  spread within crops, 12, 21, 27, 28, 31,
    44, 59, 61, 75, 81; Figs. 20, 23, 24, 25
  storage *in vitro*, 7, 79
  strains, 14, 79; T. 3
  susceptibility to infection, 12, 13, 64,
    66, 76, 79; T. 2, 3
  thermal inactivation, 1, 7, 79
  time of virus spread, 65, 67, 68, 75, 81;
    T. 16
  tolerance to infection, 12, 31, 64, 68,
    79; T. 1
Cauliflower virus 1, 1
Celosia cristata, 35
Centaurea moschata, 35
Charlock, 34, 77, 78; Pl. 7b
Cheiranthus cheiri, 12, 34
Chenopodium album, 35
C. amaranticolor, 35, 78
C. glaucum, 35
Chickweed, 35
Chinese cabbage, 1, 14, 34, 77
Chlorotic spots and rings, 9, 79; Pl. 3a,
  b, c
Cichorium endivia, 35
Cineraria, 35
Cleary, J. P., 53, 55, 84
Cockscomb, 35
Collingwood, C. A., 84
Conringia orientalis, 35
Control measures, 74, 82
Cooper, B. A., 84
Cornwall, 6, 40, 60, 75; T. 17
Cowin, S. M., 33
Crambe maritima, 34
Croxall, H. E., 7, 55, 76, 84
Cruciferae, 1, 7, 76, 78, 79, 81, 83
Cucumber mosaic virus, 1, 78, 83

Dame's violet, 35
Daniel, G. H., 84
Datchet, Bucks., 58, 60, 63, 65; Figs. 19,
  20, 21; T. 17, 25, 26, 27, 28
Datura stramonium, 78

Day-length, effect on symptoms, 16
DDT, 45, 46; T. 19
Delphinium ajacis, 35
Dennis, R. W. G., 7
Derbyshire, 12, 45, 75; T. 17
Devon, 6, 14, 17, 19, 44, 55, 61, 63, 65,
  75; T. 17, 25, 26
Digitalis purpurea, 35
Diplotaxis muralis, 34
D. tenuifolia, 34
Doncaster, J. P., 39
Duke of Argyll's tea-plant, 35
Dundee, 7, 76
Dung, 21, 80
  affecting CJM incidence, 22, 27, 28;
    T. 6, 10
  affecting cauliflower yield, 23, 28, 29,
    31; Figs. 9, 10; T. 8, 11
  affecting quality of cauliflower curd, 23,
    29, 31; Fig. 8
  affecting survival, 22, 29; Fig. 7; T. 7
  affecting time of maturity, 27
  affecting 'tip-burn', 22, 27, 28, 31; T. 7
Dunwoody, J. H. A., 84
Durham, 76; T. 17

Earwigs, 76
Edinburgh, 7, 76
Efford Experimental Horticulture Station,
  21, 27, 31, 55, 57, 58, 67, 84; T. 22,
  23, 24, 25
Empson, D. W., 40, 41, 60
'enations', 8, Pl. 1c
Endive, 35
England, 1, Figs. 1, 2, 3
  eastern, 2, 7, 41, 79, 83
  midlands, 6, 40
  northern, 6, 7, 40, 41, 83
  southern, 2, 6, 40, 41, 79
Erysimum cheiranthoides, 35
Evening scented stock, 35

Fairbank, H., 55, 84
Fat hen, 35
Fiddian, W. E. H., 84
Field penny-cress, 34
Fife, 76
Finch, C. G., 84
Flea-beetles, 1, 76, 77, 78, 83
Foister, C. E., 7, 84
Forget-me-not, 35
Forth, 7
Foxglove, 35
Fraenkel, U. E. K., 55, 84
Fungi, entomophagous, 15; Pl. 5c

'Gapping', effect on yield, 28, 29, 30, 31,
  80
Garden cress, 34
Gardens, 5, 12, 41, 76

Moreton, B. D., 40, 44, 84
Morgan, H. G., 84
Moss, N., 55, 84
Mottled leaves, 9, 16; Pl. 5 a, b
Müller, H. J., 73
Mustard, 1, 2, 9, 34, 39, 52, 53, 79; Fig. 16; T. 1, 20
Mustard beetles, 76
Myosotis palustris, 35
Myzus persicae, 36, 73, 77, 80
  incidence, 21, 27, 28, 31, 38, 52, 56, 57, 65, 68; T. 16, 17, 25
  infectivity of alatae from infected plants, 42, 81; Pl. 6a; T. 18
  overwintering, 39, 41
  retention of virus, 38
Nasturtium officinale, 2, 35

National Agricultural Advisory Service, 40, 44, 67, 84
National Institute of Agricultural Botany, 12, 28, 84
Necrotic rings and spots, 8, 9, 15, 33, 79; Pl. 1 d, 2d, 3a, 4b, c, d
Neslia paniculata, 13, 35
Nicotiana glutinosa, 35
N. rustica, 35
N. tabacum, 35
Nicotine, 45
Nitrochalk,
  affecting ClM incidence, 28
  affecting cauliflower yield, 29, 31, 80; T. 11
  affecting quality of cauliflower curd, 29, 31
  affecting survival, 29, 31
  affecting 'tip-burn', 28
Nitrogen, 21, 22, 23, 24, 27, 28, 29, 31, 80; Figs. 7, 8, 9, 10, 11; T. 6, 7, 8, 10
Nixon, H. L., 77, 78
Non-persistent viruses, 38, 50
North, C., 84
Northumberland, 55, 76; T. 17, 22, 23, 24
Nottinghamshire, 45

Ogilvie, L., 6, 7, 84
Opium poppy, 35

Papaver nudicaule, 35
P. rhoeas, 35
P. somniferum, 35
Papaver spp., 12
Parathion, 45, 46, T. 19
Peach, 41
Pepperwort, 34
Periwinkle, 35
Petunia hybrida, 12, 35
Phosphate, 21, 80
Physalis pubescens, 35
Plank, J. E. van der, 49

Potash, 21, 80
Potato, 39, 41, 50, 69
Pound, G. S., 1, 16, 33, 44, 60
Prentice, I. W., 1, 6, 7, 33, 44, 45, 61, 74, 84
Preston, N. C., 84
Protection tests, 77
Pyrethrins, 46, T. 19
Pyrolan, 46, T. 19

Radish, 34, 76
Radish virus 1, 1
Rape, 2, 34; T. 1
Raphanus raphanistrum, 34
R. sativus, 34
Reseda odorata, 35
Roberts, E. T., 58, 84
Roberts, F. M., 38
Robertson, A. G., 84
Roguing, 61, 65, 76, 82; Fig. 21; T. 28
Rosser, W. R., 84
Rothamsted Experimental Station, 9, 12, 16, 21, 28, 39, 45, 52, 53, 54, 58, 67, 68, 78, 84; Figs. 5, 6, 7, 8, 9, 10, 11, 12, 13, 14, 16; T. 1, 2, 3, 4, 5, 6, 7, 8, 9, 11, 12, 13, 14, 15, 16, 17, 19, 20, 21, 25

Salpiglossus sinuata, 35
Savoy cabbage, 2, 34, 43, 76; T. 1
Scabiosa atropurpurea, 35
Schofield, E. R., 84
Schradan, 45, 46, 53; T. 19
Schultz, E. S., 1
'Scorch', see 'tip-burn'
Scotland, 7, 76, 77, 83, 84
  Department of Agriculture, 7, 76
Seakale, 34
Seale-Hayne Agricultural College, 44, 55; T. 17, 22, 23, 25, 26
Seedbeds
  destruction of, 59, 75
  drilled and broadcast, 48, 82; Figs. 12, 13
  insecticidal treatment, 45, 53, 55, 74, 82; Fig. 12; T. 19
  isolation for virus control, 12, 44, 74, 82
  spacing and size of seedlings, 47, 52, 74, 82; Figs. 13, 14; T. 6
  time of sowing, 43, 58
  virus control by barrier crops, 50, 74, 82; Figs. 15, 16, 17, 18; T. 20, 21, 23, 24
  virus transmitted to, 39, 44, 45, 81
Senecio cruentus, 35
Severin, H. H. P., 32, 37; T. 15
Shade, effect on symptoms, 19, 20, 80
Shepherd's purse, 34, 77, 78
Shirley poppy, 35
Short, M. E., 34

Printed in the United States
By Bookmasters